HVAC Retrofits:
Energy Savings Made Easy

HVAC Retrofits:
Energy Savings Made Easy

By Herb Wendes, P.E.

Published by
THE FAIRMONT PRESS, INC.
700 Indian Trail
Lilburn, GA 30247

Library of Congress Cataloging-in-Publication Data

Wendes, Herbert.
  HVAC retrofits: energy savings made easy / by Herb Wendes.
  p. cm.
  Includes index.
  ISBN 0-88173-161-7
    1. Heating--Estimates. 2. Ventilation--Estimates. 3. Air conditioning--
Estimates. 4. Buildings--Mechanical equipment--Installation--Estimates.
5. Buildings--Energy conservation. I. Title.
TH7335.H46   1994   697--dc20   93-45886
                                                                    CIP

Published by The Fairmont Press, Inc.
700 Indian Trail
Lilburn, GA 30247

Printed in the United States of America

10 9 8 7 6 5 4 3 2 1

ISBN 0-88173-161-7   FP

ISBN 0-13-103748-X   PH

Distributed by PTR Prentice Hall
Prentice-Hall, Inc.
A Paramount Communications Company
Englewood Cliffs, NJ 07632

Prentice-Hall International (UK) Limited, London
Prentice-Hall of Australia Pty. Limited, Sydney
Prentice-Hall Canada Inc., Toronto
Prentice-Hall Hispanoamericana, S.A., Mexico
Prentice-Hall of India Private Limited, New Delhi
Prentice-Hall of Japan, Inc., Tokyo
Simon & Schuster Asia Pte. Ltd., Singapore
Editora Prentice-Hall do Brasil, Ltda., Rio de Janeiro

Table of Contents

PART I
GENERAL STEPS IN PLANNING AND BUDGETING AN HVAC ENERGY RETROFIT PROGRAM

Section 1

ENERGY SAVINGS TECHNIQUES

Section 2
AUDITING

Section 3

ENGINEERING AND FINANCIAL EVALUATIONS

Section 4

ESTIMATING RETROFIT COSTS

Section 5

REDUCE HEATING AND COOLING LOADS

Section 6

REDUCE FLOWS AND RESISTANCE

Section 7

HVAC SYSTEM CONVERSIONS

Section 8

INDOOR AIR QUALITY

PART II
How To Estimate and Retrofit
Specific HVAC System Components

Section 9

HEATING AND COOLING EQUIPMENT

Section 10

HVAC UNITS AND AIR DISTRIBUTION EQUIPMENT

Section 11

DUCTWORK, PIPING AND INSULATION

Section 12

CONTROLS AND ELECTRICAL

Section 13

HEAT RECOVERY

Section 14

TESTING, GENERAL CONSTRUCTION WORK AND MISC.

Section 15

APPENDICES

CHARTS AND FORMULAS

Preface

There is no investment that pays a better rate of return with less risk than an HVAC energy conservation investment.

HVAC Retrofits: Savings Made Easy, covers in a simplified fashion all calculating costs of labor, material, equipment and subcontracts for retrofitting all types of HVAC equipment and systems and how to calculate the resultant energy savings.

It points the way to minimizing retrofit costs and maximizing energy savings. It shows you how to achieve faster paybacks and higher rates of return.

HVAC Retrofits: Savings Made Easy, covers easy step-by-step procedures on how to convert constant volume systems to variable air volume and how to accurately calculate the prices of doing so.

It shows you how to estimate labor and material costs on all types of HVAC equipment and systems:

- Air Distribution Equipment
- HVAC Units
- Heating Equipment
- Cooling Equipment
- Ductwork and Accessories
- Piping and Accessories
- Energy Management Systems
- Controls
- Insulation
- Lighting and Electrical Work
- Energy Recovery Equipment

It covers how to estimate labor and material costs for:

- Auditing & Engineering
- Converting to VAV (Variable Air Volume)
- Reducing Building Heating and Cooling Loads
- Reducing Fan and Pump Volumes
- Reducing Ductwork and Piping Resistance
- Remodeling Costs
- Balancing and Monitoring Costs

What's involved in the entire spectrum of an energy retrofit program from the initial walk-through audit to the monitoring after the retrofit is complete is covered. Step-by-step procedures guide you through the preparation of retrofit estimates.

HVAC Retrofits: Savings Made Easy tells you what types of energy auditing and monitoring instruments you should use, when and where to purchase them and what the costs are.

The book provides conceptual pricing for feasibility purposes, engineering budgets for budgeting projections and detailed labor and material costs for fully detailed estimates for firm bids.

The book is written for people involved with HVAC systems and saving money on energy:

- Building Engineers
- HVAC Contractors
- Building Maintenance Departments
- Owners, Managers
- General Contractors and Builders
- Control Contractors
- Electrical and Plumbing Contractors
- Design Engineers and Architects

It applies to all types of buildings:

- Hospitals
- Schools
- Office Buildings
- Factories and Plants
- Restaurants
- Transportation Facilities
- Utilities
- Stores
- Government Buildings

Special Features

This book covers how to calculate, analyze and project energy consumption in terms of Btu's and kW's and in cost terms.

Sample energy retrofit estimates are provided on various types of buildings and systems.

A comprehensive, handy reference list of virtually all energy saving techniques and ideas are included. There are listings of energy conserving methods for retrofitting different types of buildings and for different types of systems.

How to derate fan and pump volumes commensurate with reduced building heating and cooling loads to save operating costs is covered.

You make a walk-through inspection of a typical hospital, school, office building, store or plant.

How to sell audits and retrofits either to management or potential customers is covered.

Part 1

General Steps in Planning and Budgeting an HVAC Energy Retrofit Program

Section 1

HVAC Energy Savings

Chapter 1, Introduction To HVAC Retrofitting

Chapter 2, Energy Savings Techniques

Chapter 1

Introduction To
HVAC Energy Retrofit Work

You make a walk-through inspection of a typical hospital, school, office building, store or plant.

You find a fan is pumping out 40% more air than is needed, wasting electrical energy.

You find a pump is also delivering an excessive amount of water, again wasting electrical energy.

Air is being cooled to 55°F at the central HVAC equipment and then reheated at the terminal unit in the spaces in a constant volume reheat system.

A chiller is nearly twice the size it needs to be.

Fifty percent more minimum outside air is being drawn in the building than is really needed because controls are not set right or linkages etc. malfunctioning.

The HEPA filters for the operating rooms are dirty and the pressure drop of the system is double of what it should be and the air delivery way down.

As you inspect more and more systems, equipment and areas of the building, the energy waste goes on and on.

The oversized chiller cycles and runs at a low level of efficiency in terms of kW/tons. You ask yourself what are the various energy saving

5

options available to reduce this waste and loss. You decide to consider replacing the chiller and a multitude of questions flood upon you.

How much does it cost to replace the chiller with a new correctly sized one, an energy efficient one with the latest control features?

MANY QUESTIONS AND PROBLEMS NEED TO BE RESOLVED

What does it cost? What materials are required? What labor is involved for any energy retrofit improvement? Many questions must be answered and prices calculated regardless of what the retrofit involves.

1. What various options are available to rectify this waste, what is the retrofit cost of each and how much will each save in energy costs?

2. What is a realistic purchase price of any new equipment needed?

3. How much labor is needed to remove the old one and install the new one?

4. What piping and valve changes will be required?

5. What ductwork changes will be needed?

6. What electrical wiring and parts costs are there?

7. What temperature control costs will there be?

8. What will hoisting and material handling costs be?

9. If the pump or fan can be changed, what costs are involved? If they can't be used, what are the costs in removal and replacement?

10. What will the costs of insulation be?

11. How much labor will be involved in draining original system, flushing, pressuring, testing and refilling new system and start up?

12. How much will be needed for balancing and adjusting the system and monitoring energy costs?

13. And lastly, the big question, what will the energy savings be with this approach and what is the payback and return on investment?

It is absolutely necessary to obtain this information and compare the various avenues available and make a wise decision based on accurate and thorough cost projections and energy savings.

SAMPLE ENERGY SAVINGS

If you reduce the volume on an over-capacity fan or pump 25 percent the horsepower is reduced 57 percent.

At 10 cents per kWh it costs $1000 per yr. per BHP drawn on a fan or pump running full time.

It costs $500 per yr. per running half time.

Therefore, a 20 HP motor costing $20,000 per yr. to run full time if reduced to a 10 BHP load can save $10,000 per year.

The payback then in this case would be between .1 and .2 years on a full time basis and between .2 and .4 years on a half time basis.

ENERGY RETROFIT WORK SPANS ALL DISCIPLINES

You will find that estimating energy retrofit costs and savings covers all the HVAC disciplines, engineering, piping, sheet metal, temperature controls, insulation, electrical, computers, etc. maintenance, service, building management.

This manual provides you with explanations of how to perform various retrofit changes, how to calculate the energy savings and the renovation costs. It provides procedures and formulas for energy programs, audits, engineering and estimating to help you through this maze of complexity and data.

For Each Dollar Wasted in Energy Usage It Takes $20 More in Gross Sales to Recuperate It at A Profit Margin of 5 Percent.

THE ENERGY SAVINGS GAME

Overall Procedure For An Energy Management Program

The overall procedure for an effective energy management program is as follows:

Phase 1

1. The first step is a decision to pursue some kind of energy savings program, a decision to check out waste and evaluate potential energy savings.

2. The second step in an energy management program is phase one of the energy audit, the walk through portion. This involves the initial examination of the energy consumption history of the building, gathering information on the building, systems and its equipment, noting operation and maintenance conditions and problems, listing apparent energy waste problems and obvious potential energy savings improvements to this point.

Phase 2

3. Phase two of the energy audit involves field tests of flows, pressures, temperatures, etc., further field inspections, getting name plate data and in depth interviews with building personnel. Phase two further involves determining the required heating, cooling and Cfm loads of various areas of the building, determining the energy consumption of specific equipment and systems, a detailed listing of energy problems systems and a complete building and systems report.

4. Generate and develop potential energy savings improvements, operation and maintenance corrections, reducing heating and cooling loads, reducing flows and resistances of HVAC systems, considering more energy efficient equipment and systems, lighting, electrical, controls, heat recovery possibilities, solar, etc.

5. Calculate the potential energy savings of the various improvements.

6. Estimate the retrofit costs involved.

7. Evaluate paybacks and return on investments.

8. Select with the owner which improvements to proceed with and

assign priorities based on maximum savings, minimum investments, feasibility, etc.

9. Properly engineer prepare drawings, specification, etc. as required.

10. Obtain quotations, award contracts and proceed with the renovation work.

11. Start-up renovated systems properly and test and balance.

12. Monitor systems and equipment. Modify as required.

START ▶	DECISION TO PURSUE ENERGY SAVINGS PROGRAM	WALK THRU AUDIT	IN DEPTH AUDIT	LIST OF PROBLEMS	JAIL
FINISH: COLLECT SAVINGS					GENERATE ENERGY IMPROVEMENTS
MONITORING, MODIFICATIONS		**ENERGY SAVINGS**			IMPROVE OPERATION & MAINTENANCE
START-UP, TESTING & BALANCING		**GAME**			REDUCE HEAT/COOL LOADS
QUOTATIONS, AWARDS, CONSTRUCTION					REDUCE FLOWS & RESISTANCE
DESIGN, DRAWINGS, SPECIFICATIONS		CHANCE			IMPROVE HVAC SYSTEM & EQUIPMENT EFFICIENCY
SELECT IMPROVEMENTS, ASSIGN PRIORITIES					LIGHTING, ELECTRICAL, MOTORS
GO TO JAIL	EVALUATE PAYBACK	ESTIMATE RETROFIT COSTS	CALCULATE ENERGY SAVINGS	HEAT RECOVERY SOLAR	CONTROLS EMS

Chapter 2

Energy Savings Technique

This chapter contains lists of methods and ideas which can be used for saving energy in HVAC systems.

The 14 commandments of HVAC energy conservation are an excellent outline of the realm of conservation and the individual lists on reducing loads, heating and cooling energy reductions etc. Cover each area in a bit more detail.

It is important to guide your thinking about energy conservation in HVAC systems in the following manner.

THINKING ABOUT ENERGY CONSERVATION

1. **Generalities:** Start off energy conservation programs thinking in terms of principles or generalities and then follow up with particulars. Think about reducing HVAC loads, O & M savings, improving efficiency of equipment and systems, reducing flows, etc. The 'Commandments and Principles of Energy Conservation are covered in this chapter.

2. **Specifics:** After a general concept is formed then think in terms of specific heating and cooling equipment, particular HVAC systems, piping systems, ductwork systems, insulation, controls, etc.

3. **Load Variations:** Think about how the building heating and cooling loads may vary due to occupancy, the shifting sun, operations etc. and about which heating and cooling loads are constant on a daily basis without variation.

4. **Low Cost, No Cost Items:** Think in terms of no cost, low cost energy saving measures which can be done easily and quickly and which may have phenomenal pay backs.

- Changing space thermostat settings
- Reducing lighting
- Turning lighting off during non-occupied times
- Turning HVAC equipment off when not occupied
- Reducing minimum outside air
- Reducing flows
- Using the free outside air for cooling
- Up-to date maintenance
- Improving combustion
- Reducing ductwork and piping resistance
- Changing air and water supply and return temperatures
- Employing temperature resets on HVAC systems
- Test and Balance

5. **Capital Investment Items:** Then think in terms of capital investment energy improvements searching for those with the greatest energy savings and the highest rate of return or fastest payback.

6. **Electrical:** Distinguish electrical consumption costs of fans, pumps, chillers, condensers, cooling towers, lighting etc. from fuel consumption items such as boilers, burners etc.

THE 14 PRINCIPLES OF ENERGY CONSERVATION

I   Keep It Maintained
- Clean It
- Seal It
- Adjust It
- Lubricate It
- Keep It Unobstructed
- Recalibrate
- Check Control Settings
- Check Speeds
- Check Flows
- Keep Valves, Dampers, Etc. Operating Smoothly

II   Reduce Heating And Cooling Loads
- Reduce Space Temperature Settings in Winter, Increase in Summer
- Reduce Minimum Outside Air

- Reduce Lighting Loads
- Provide Maximum Economical Insulation
- Double or Triple Glaze Windows
- Seal Building Infiltration and Exfiltration Leaks
- Provide window Protection Against Solar Radiation
- Reduce Absortivity of Walls and Roof With Lighter Colors
- Control Temperature Stratification of Air in Spaces

III **Turn It Off When Not In Use**
Turn On/Off At Optimum Times

IV **Reduce Flows In Air And Hydronic Systems**
Reduce Resistances In Air And Hydronic Systems

V **Use More Efficient Equipment**
- Higher SEER'S
- Lower kW's Per Ton
- Higher Seasonal C.O.P.'s
- Lower Pressure Drop
- Avoid Over Sized Equipment
- Minimize Cycling. Use Correct Size Equipment
- Use Staging of Compressors, Boilers, Coils, Etc.

VI **Improve Efficiencies of Systems**
- Use Variable Air Volume
- Use Variable Water Volume
- Avoid Simultaneous Heating and Cooling
- Use Temperature Resets

VII **Improve, Update Controls**
- Sensors
- Control Motors
- Controllers
- Stats
- Valves, Dampers

VIII **Use Computer Controlled Equipment**
- EMS Control Units
- EMS Modular Units
- Dedicated Equipment Micro Processors
- Timers
- Set Backs, Set Ups

IX   Recover Waste Heat

X   Use Natures Free Energy
- Outside Air For Cooling
- Ground Water
- Solar Heating

XI   Test And Balance Systems For Maximum Efficiency

XII   Perform Valid Audit As Basis of Retrofit Before Starting Energy Saving Measurers
- Obtain Thorough and Accurate Information on Actual Consumption, Systems, Operations and Conditions Before
- Make Thorough and Reliable Engineering and Costs Calculations Before
- Considerate Various Alternates

XIII   Start With No Cost And Low Cost Energy Savings Measures

IVX   Monitor, Adjust, Modify As Needed

REDUCE HEATING AND COOLING LOADS

Change Space Temperature Settings
- Turn thermostats down to 70°F in winter.
- Turn thermostats up to 75°F in summer.
- Set stats back or set about 10°F evenings and weekends.
- Do manually, with individual set back stats or with E.M.S.

Reduce Amount of Lighting
- Reduce lighting levels. Delamp.

Reduce Minimum Outside Air Quantities

Turn HVAC Equipment Off When Not Occupied
- Shut down HVAC equipment during non-occupied hours, nights, weekends, etc. Put into effect on a manual basis or install small modular E.M.S. units.

- Turn domestic hot water pumps off during non-occupied hours, nights, weekends, etc.

- Turn off all toilet, kitchen and lab exhaust fans during non-occu-pancy, nights, weekends.

- Shut down make-up air units and HVAC zones when unoccu-pied.

Turn Lighting Off When Not Occupied
- Turn off lighting during non-occupied hours, nights, weekends, etc. Execute on a manual basis or install timers or small modular E.M.S. units.

- Shade windows during summer.

- Draw drapes, curtains, blinds to reduce solar loads during sum-mer.

Turn Heat Producing Equipment Off When Not In Use

Reduce Structural Transmission, Solar And Infiltration Gains And Losses
- Provide maximum economical insulation.
- Double or triple glaze windows.
- Seal infiltration leaks.
- Provide window protection against solar radiation summer.

Use Free Energy
- Use outdoor air for free cooling when O.A. temperatures are roughly between 40°F and 60°F during cooling season.

- Use outdoor air for free cooling during winter when outside air temperatures are below 60°F in spaces that require cooling year round.

- Where economizers exist, make sure they operate properly.

Reduce Absortivity of Walls And Roof With Lighter Colors

Control Temperature Stratification of Air In Spaces With Ceiling Pro-peller Fans, Etc.

KEEP IT MAINTAINED

HVAC Equipment Maintenance
- Keep filters, strainers, traps, coils, condenser tubes, boiler tubes etc. clean to minimize resistance and maximize heat transfer.

- Check and adjust flows, rpm's, amp draws of fans, pumps, chillers and adjust for maximum efficiency.

- Make sure everything is sealed.

- Make sure everything is lubricated.

- Check drives for proper tension alignment, wear, correct power transmission, etc.

Controls Maintenance
- Check calibrations on thermostat, aquastats, pressure stats, etc. and correct as required.

- Check operation of control motors for dampers and valves to see that they operate correctly through full cycle, the linkages are fastened correctly and make sure that there is no binding either in the damper or valves as well as with the linkages.

- Make sure economizer cycles operate properly.

REDUCE FLOWS AND RESISTANCE

Reduce Fan Volumes
- Reduce fan volume flows commensurate with reduced heating and cooling loads.

- Reduce exhaust fan air volumes where feasible.

Reduce Pump Volumes

- Reduce pump flows commensurate with reduced heating and cooling loads and/or due to over design.

Reduce Resistance of Air Distribution Systems

Reduce Resistance of Piping Systems

IMPROVE EQUIPMENT EFFICIENCY

- High SEER's (Seasonal Energy Efficiency)
- Lower kW's per ton for chillers
- Higher seasonal C.O.P.'s (Coefficient of Performance)
- Lower equipment pressure drops.
- Keep combustion efficiencies at peaks.
- Keep refrigerants at efficient levels.
- Stage electric heating coils.
- Stage burners.
- Stage compressors.
- Avoid oversized equipment.
- Minimize cycling.

HEATING ENERGY IMPROVEMENTS

Combustion Efficiency
- Check combustion efficiency and adjust. Read stack temperature and CO_2 or O_2 and tune up. Reduce stack temperatures and minimize excess air.

- Install automatic oxygen trim controller to maintain maximum combustion efficiency at all times.

- Clean soot, scale etc. from tubes and firewalls.

- Schedule boiler blow down on as-required basis rather than on a regular schedule.

- Check boiler areas for building negative pressure. Correct to provide proper amount of combustion air.

Staging Heating
- First burner on at 50°F outside.
- Second burner on at 30°F outside.

- Third burner on at 10°F outside.

Reset Temperatures
- Reset boiler hot water discharge temperature proportional to outside air temperature. If the outdoor temperature is 50°F or 60°F lower hot water temperature to 120°F. If outdoor temperatures are 0°F or 20°F raise temperature to 160°F.

- Install automatic temperature reset controls.

Vents, Drafts
- Install automatic draft or vent damper to reduce stack losses when boiler is not firing.

Multiple And New Boilers
- Install either multiple boilers in place of one operating excessively at part load or a single smaller boiler.

- Operate one boiler to its maximum load before bringing other boilers on line.

Insulation
- Insulate boilers, HVAC units, reduce radiation losses with burners etc. which are in unheated spaces, outside on roof or in air conditioned spaces.

Heat Recovery Equipment
- If stack temperatures exceed 300°F install a heat recovery unit in the stack and use reclaimed heat to heat up make up air, combustion air, domestic hot water or for space heating.

- Install turbulators or baffles to improve heat transfer if stack temperatures are too high (over 450°F) and combustion efficiency is at a maximum.

Burner And Type Fuel
- Convert to more efficient fuel
- Install new, more energy efficient burners
- Install electronic ignitions in place of standing pilot lights.

Feedwater
- Install an economizer for preheating boiler feed water.
- Maintain proper level of additives and chemicals.

Steam
- Clean steam traps

- Seal leaky steam traps

- Monitor steam trap leakage

- Keep steam pressure as low as feasibly possible for coils or radiation equipment.

- Vary steam pressure proportional with demand.

- Install steam pressure regulators to reduce steam usage. Use by passes only for repairs and emergencies.

- Install steam flow meter on larger systems.

- Install heat recovery units in condensate pump systems to reclaim heat and reduce temperature of the condensate.

- Use blowdown heat recovery

Oil
- Install automatic viscosity controls on fuel oil systems for maximum atomization and the ability to use other grades of fuel etc.

- Reduce firing rates of oil or gas burners if boiler is oversized to avoid short cycling.

COOLING ENERGY IMPROVEMENTS

Chillers
- Raise chilled water temperature leaving evaporator 2°F.

- Clean condenser tubes yearly or install automatic tube cleaning system.

- Change one large chiller operating at an inefficient partial load to two or more chillers staged to operate at more efficient full loads in sequence.

- Replace older, inefficient chillers at high C.O.P.'s such as .9, 1.0, 1.1 kW/ton to more efficient chillers in .6, .7 kW/ton range.

- Replace oversized chillers running inefficiently and cycling with new high efficiency models.

- Steam absorption chillers need steam the year round and are less efficient. Two-stage absorption chiller uses 30% to 40% less energy than single-stage uses high-pressure steam. With the use of steam, the cooling tower can be smaller, helping reduce the life-cycle cost of the HVAC system.

- Install double bundle reheat condensers in buildings that have very large interior areas and a lot of heat to recover. Even in cold weather heat can be recaptured.

- Install variable volume chilled water pumping system.

- Use efficient staging. Operate one of several compressors at full load before starting second.

- Use multiple staging with flash intercooling.

- Install controls for multiple chillers so that one chiller can run at full capacity before second is activated.

- Increased heat exchanger surface areas increased efficiency.

- Install variable speed drive on chillers compressors.

- Replace existing compressor and motor only to match reduced load and retain exchangers.

- Install open compressor motors on chillers versus hermetic motors.

- Use thermocycle.

- Use strainer cycle.

- Use outdoor air for cooling in mild weather without running any refrigeration equipment.

- Convert dual duct mixing box, terminal reheat and multi-zone systems to systems which don't require simultaneous heat and cooling thereby being able to shut the cooling system in winter.

Cooling Towers

- Lower cooling tower condenser water temperature returning to chiller.

- Use cooling tower water temperature reset.

- Ceramic cooling tower offers life-cycle benefits of 100% thermal performances for over 25 years uses less electricity and needs less maintenance.

Refrigerant Systems

- Reduce head pressure.

- Adjust suction pressure on DX air conditioning system to raise refrigerant suction temperature 4°F.

- Install hot gas heat exchanger to recovery heat.

- Stage condensers:
    First one on at 60°F outside
    Second one on at 80°F outside

- Obtain free cooling by using an auxiliary heat exchanger connected in parallel to the chilled water supply and return piping between the chiller and the cooling coils. See ASHRAE manual.

- Obtain free cooling by installing water circuit interconnection loop in chilled water supply piping before cooling coils.

IMPROVE HVAC SYSTEM EFFICIENCY

Automatic Resettings of Temperatures

On systems that heat and cool simultaneously, such as terminal reheat, dual duct and multi-zone, savings can be achieved by resetting cold and hot deck supply temps as a function of space temps.

Low Pressure, Constant Air Volume Systems
Multi-Zone Systems

- Lower the hot deck discharge air temperature on low pressure multi-zone system as close as possible to meet the demand of the zone with the most heating demand.

- Raise the cold deck discharge air temperature as close as possible to meet demand with the most cooling demand.

- Use a load analyzer which compares the temperature of each zone in both of the above cases.

Terminal Reheat Systems

- Constant air volume terminal reheat systems with fixed 55°F supply air temperatures waste 30 to 60 percent of the heating energy. The waste is caused when precooled air is reheated at the reheat coil and when preheated air is used for cooling. If every zone is reheating the cold deck temperature is too cold and both cooling and heating energy are being wasted.

- Raise cold deck supply air temperature to 58°F in low pressure, single duct terminal reheat systems so that a minimal number of reheat coils are activated.

- Reset cold deck temperatures higher by installing zone load analyzers which measure the demand of each zone, set the cold deck temperatures up and minimize amount of reheat required.

Medium And High Pressure, Constant Air Volume Systems
Terminal Reheat Systems

- Raise cold deck supply air temperature on pressure high pressure terminal reheat systems.

Induction Systems

- Shut off perimeter induction system fans in winter and use induction units as baseboard heating. Run fan at reduced rate during summer and don't run at all in winter.

Dual Duct Systems

- Lower hot duct temperature and raise cold duct temperature in dual duct systems so that most important loads in building can just be met.

- When there is no heating or cooling load required in the dual duct system between seasons, operate on cold air duct only.

- With dual dust operation in winter, shut down the cold air duct, chillers and chilled water pumps and in summer, shut down heating equipment and hot duct.

- Reduce supply fan flow to a minimal acceptable level according to actual current needs.

- With computerized controls reset hot duct temperatures according to the velocity pressure in the hot deck and reset cold duct temperatures according to the velocity pressure in the cold deck.

- If the dual duct system only serves interior areas which require cooling the year round, shut the hot duct and heating coils down permanently and just run on cooling as single duct constant air volume system.

- Change system to single duct operation, but still using both the hot and cold ducts. Close the hot duct in summer and shut off all the heating. Close the cold duct in winter and shut off all the cooling such as pumps, chillers, etc.

- Convert system to single duct variable air volume system by shutting hot duct down permanently.

- Keep the dual duct system and mixing boxes but convert to both cold duct variable air volume and hot duct variable air volume. This is done by putting separate VAV control motors on both the hot and cold duct dampers in the mixing box. The operation is such that neither the hot or cold damper will open unless the other is 100 percent closed.

LIGHTING AND ELECTRICAL ENERGY SAVINGS

Reduce Lighting Levels
- Reduce lighting wattage per sq ft of building commensurate with actual current needs.

- Remove selected florescent bulbs and disconnect or cut wires or re-move ballasts.

- Completely disconnect selected fixtures and remove.

  Electric lighting generates 60 percent more heat per unit of illumina-tion than sunlight thus requiring more cooling expenditures. This should be evaluated when considering less lighting and more natural light against increased conduction and solar heat gains through win-dows.

Replace Lighting
- Replace less efficient incandescent lighting with more efficient flores-cent fixtures.

- Install high efficiency lighting such as high pressure sodium.

- Install higher efficiency florescent bulbs.

- Install higher efficiency ballasts.

- Install electronic ballasts.

Operation Maintenance
- Turn lights off when spaces are not occupied.

- Clean bulbs, reflectors and lenses yearly.

- When natural light is available in perimeter areas or through skylights etc., turn off lights.

Efficient Lighting System Design
- Use specific task lighting.

- In high ceiling areas, lower lighting fixtures.

- Lower ceilings.

- Split up lighting controls so one control does not operate all lights in larger areas for local control of occupied and unoccupied spaces.

- Install mirror surface reflectors inserts in florescent fixtures for high efficiency lighting. Can save up to 50 percent on electrical costs and lamp replacements, plus up to 25 percent of air conditioning costs.

Outside Lighting
- Reduce outside lighting commensurate with safety.
- Use photo cell control on outdoor lights.

Motors
- Install smaller motor commensurate with actual load so that existing motor doesn't run on partial load causing less efficiency and lower power factor.

- Install more energy efficient motors.

- Install variable speed variable frequency inverter motors.

Starters
- Install solid state soft motor starters.

RECOVER WASTE HEAT

How To Use Recaptured Heat
1. Heat up makeup air or temper outside air for HVAC systems.

2. Preheat combustion air.

3. Preheat hot water for central heating coils or terminal reheat coils in HVAC units.

4. Preheat hot water for baseboard heating.

5. Use heat recovery air directly for space heating or mix with regular supply air.

6. Preheat domestic hot water.

Stack Heat Recovery
- Install heat recovery unit in boiler stacks.
- Install heat recovery unit in incinerator stacks.

Engine Exhaust Heat Recovery

- Capture heat from engine exhaust.

Hot Exhaust Air Heat Recovery
- Install heat recovery units in kitchen exhaust systems, over baking ovens.

- Capture heat from PAINTING PROCESSES, paint spray booths, parts washers, parts dryers and paint baking ovens.

- Install heat recovery units in printing plants.

- Install heat recovery units in industrial chemical exhaust systems.

- Recover heat from industrial drying exhausts of grain, milk, coffee, etc.

- Recapture heat from ovens used for baking, drying, etc., from kilns.

- Install heat recovery units in heat treating plants and annealing furnaces.

- Capture heat from clothes dryers in laundries.
- Install heat recovery equipment in foundries.

Water Cooled Condenser Heat Recovery
1. Reclaim heat from chiller with double bundle condenser which would otherwise be dumped to the atmosphere via the cooling tower.

2. Recapture heat with twin tower enthalpy recovery system.

Air Cooled Condenser Heat Recovery
- Reclaim heat from hot condenser air from air cooled DX condenser.

- Reclaim heat from hot refrigerant gas with gas to air or gas to glycol water transfer in ice skating rinks, supermarkets, computer rooms etc. Use for hot water heating or space heating.

Recirculate Warm Air From Internal Heat Gains
- Recirculate heat from equipment in computer rooms.

- Recirculate heat from interior lighting fixtures to heat perimeter areas.

Steam
- Recover heat from steam condensate.

Domestic Hot Water Recovery
- Reclaim heat from domestic hot water.

Machinery Heat Recovery
- Reclaim heat from water used to cool machinery.

DOMESTIC WATER SAVINGS

- If the hot or cold water pressures are greater than 40 PSI in low rise buildings, use pressure reducing valves to lower the pressures.

- Use water restrictors in faucets.

- Lower domestic hot water temperature to about 110°F for washing hands etc.

- If one hot water heater can handle the hot water load, where there are multiple units, just run one.

- Lower laundry room hot water to 160°F.

- Where hot water is circulated with house pumps turn off pumps in unoccupied areas.

Section 2

Auditing

Chapter 3

HVAC
Energy Auditing Procedures

The purpose of an energy audit is to determine the energy consumption and costs of the overall building and of its specific components, the structure, systems and equipment. It is to generate energy improvement options, to project energy savings, to estimate the costs of energy improvements, calculate paybacks, and on this basis evaluate the various options.

A good audit is diagnostic in nature, develops a valid prognosis of the causes of energy wastes, and leads to scientific established remedies.

There are two basic phases or types of audits, short walk through audits and in depth detailed audits, either of the entire building or of only selected parts of building.

PHASE ONE: WALK-THROUGH ENERGY AUDIT PROCEDURE

1. Make an initial walk-through inspection to become familiar with the building, systems, equipment, maintenance, operation status, etc.
2. Study the plans and specifications and become familiar with the building, systems, capacities, equipment, etc.
3. Talk briefly with the building operating personnel, owner, occupants etc. about HVAC systems, comfort, problems, etc.
4. Examine the overall building energy consumption history from the owner if available. If not, get a complete energy consumption history on gas, oil, electrical, etc., from utility companies and fuel suppliers. Compare the Btu consumption per sq.ft. per year with other similar buildings and determine degree of variance.
5. List maintenance, cleaning, adjustment, repairs and balancing needed to this point. Determine what maintenance and repairs must

be done before the detailed audit can be performed.

6. Take some spot test reading if needed.

7. If a more extensive audit is needed, determine what test readings, inspections, analysis, calculation, etc. are required and estimate the time and costs involved.

8. Fill out a "Building and Systems Description" report.

9. Write out a list of existing energy problems.

10. List obvious and potential energy savings improvements. Develop the most promising energy improvements further.

11. If the walk-through audit is sufficient, calculate energy savings for the various energy improvements, estimate retrofit costs and calculate paybacks.

12. Select with the owner which energy improvements to proceed with and assign priorities. Properly engineer retrofit work and proceed.

Chart 3-1

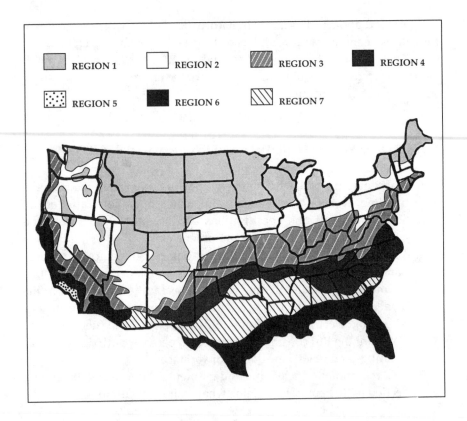

PHASE TWO: IN-DEPTH ENERGY AUDIT PROCEDURE

Field Surveys

1. Make a thorough inspection of building systems and equipment and become thoroughly familiar with them. Check out operations, performance, maintenance, malfunctions, comfort, problems, etc.

2. Check name plate data on equipment.

3. Conduct in-depth interviews with building personnel. Review maintenance, scheduling, performance, comfort and problems of building, equipment and systems.

4. Become familiar with actual hours of operation of systems and equipment, and the hours of occupancy of the personnel.

Energy History

5. Study and analyze at 3 year history of the buildings electrical and fuel energy consumption. Compare with building consumption indexes of similar buildings.

Field Tests

6. Take test readings of actual flows, temperatures, pressures, rpm's, amps, volts, etc., at HVAC equipment.

   Check pressure drops across filters, coils, strainers, etc. Check outside air flows at minimum and maximum.

   Monitor readings over a period of time with a recording instrument where required.

   Check lighting levels.

Seasonal and Peak Energy Calculations

7. Determine the actual existing seasonal and peak energy consumption and efficiencies of specific systems and equipment, etc. based on tests and other data.

8. Calculate the peak and seasonal heating, cooling and Cfm loads actually required for to meet current conditions for the overall building and various areas of the building. Compare with the design and existing capacities.

Evaluation of Energy Improvements

9. List all problems with building, systems and equipment.

10. Generate energy improvements and develop those with most potential. Write out list of improvements.

11. Calculate the potential energy savings in terms of Btu's and kWh's, and in costs.

12. Estimate costs of retrofit work.

13. Calculate paybacks and return on investments.

Review And Decisions

14. Review with owner or his representatives:

    | Problems | Costs of improvements |
    | Energy improvement options | Payback, return on investment |
    | Potential savings | |

    Consider a change only on one portion of the recommended energy improvements to test and validate the savings and to observe the effects.

15. Select with the owner which energy improvements to proceed with and assign priorities.

Engineering And Construction

16. Property engineer the owner retrofit work, prepare drawings and write specifications.

17. Obtain quotations, award contracts and proceed with the retrofit work.

18. Monitor units of energy and costs savings after put in operation. Make adjustments and modifications as required.

Chart 3-2
In-Depth Energy Audit Procedure

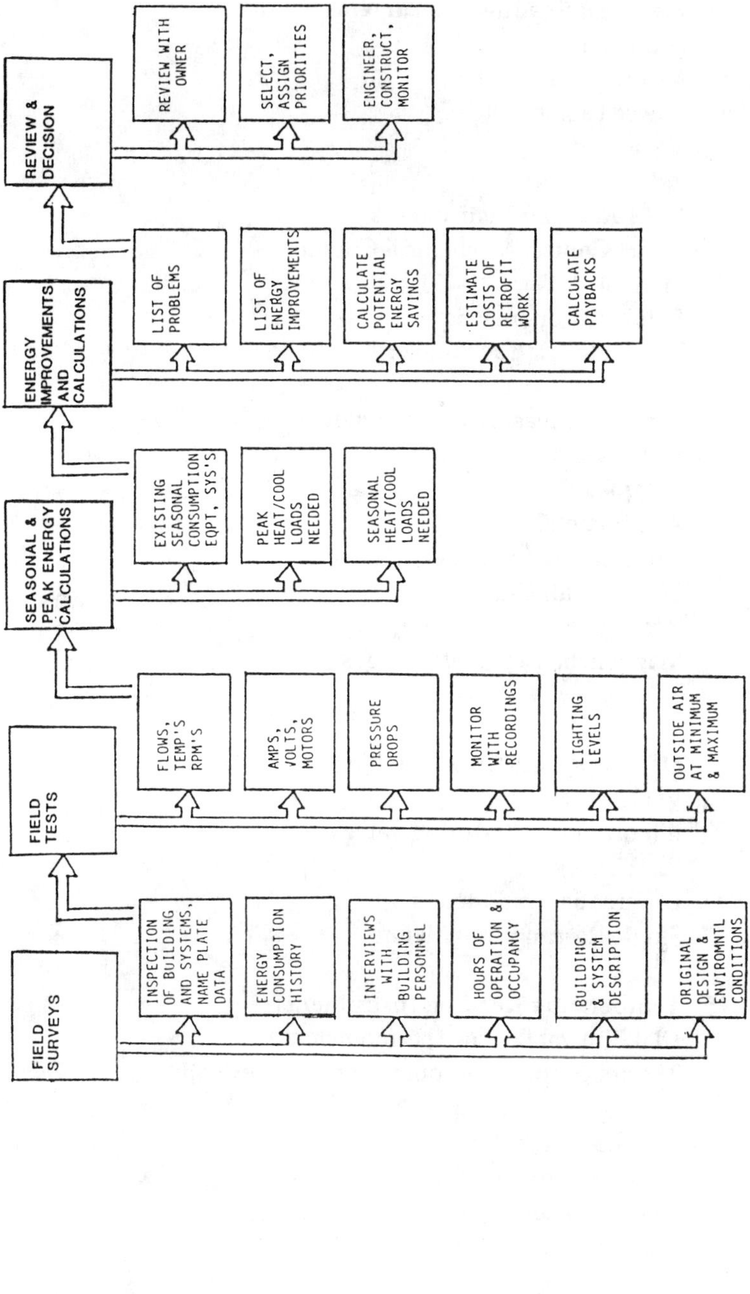

Table 3-1
Energy Auditing Instruments

1. **Electrical Reading Instruments**
   Ammeter
   Recording Ammeter
   Power Factor Meter
   kW Meter

2. **RPM Reading Instruments**
   Direct Contact Mechanical Counter
   Manual
   Biddle Chronometric
   Photo Electric Tachometer

3. **Air Flow Measuring Instruments**
   Flow Hood
   Velometer
   Anemometer
   Hot Wire Anemometer
   Liquid Draft Gauges
   Pitot Tube
   Magnehelic Dry Draft Gauges

4. **Water Flow Measuring Devices**
   Orifice Devices Circuit Setters
   Venturis
   Illinois Flow Measuring valves
   Pitot Tubes
   Differential Dial Meters
   Single Reading Dial Meters

5. **Temperature Reading Instruments**
   Dial Bimetal Pocket Thermometers
   Thermocouple Electronic Thermometer with
       Single and Multiple Probes
   Surface Pyrometer
   Temperature Recorder
   Psychrometer
   Humidity Recorder

6. **Light Reading Instruments**
   Footcandle Meter

7. **Heat Scanners**
   Infra Red Scanner
   Remote Surface Temperature Scanner and Btu Per Hour Meter

8. **Combustion Testing Kit**
   Stack Thermometer
   Oxygen Reading Gauge
   Quarter Inch Draft Gauge

9. **Refrigeration Instruments**
   Manifold Gauge
   DX Leak Detector

10. **Personal Items**
    Hand Tools
    Flash Light
    Calculator
    Watch
    Tape Measure
    Tape Recorder With Built in Mike
    Camera
    Binoculars

Each HVAC system is somewhat unique and its particular characteristics can only be identified by inspection and measurement. Information required to understand the present operation of a system and to provide a basis for deciding which modifications are likely to prove beneficial is tabulated in Table 3-2. Measurement should be as nearly simultaneous as possible.

Table 3-2
Readings Required for Thorough Audit of System

Electrical Readings	(Amps, Volts, Power Factors, kW) Fans Pumps Compressors Condensers Chillers Lights Owners operating equipment
Air Flow Rates	Total supply air from fan Total return air to fan Total outdoor air Trunk ducts Terminal units Air cooled condenser
Air Pressure Readings	Suction and discharge of fans Drops across coils, filters etc.
Water Flow Rates	From pumps Through boilers Through chillers Cooling towers Heat exchangers Coils and terminal units
Water Pressure Readings	Suction and discharge of pumps Drops across strainers, coils etc. Drops across boilers, chillers, condensers
Temperatures, Air	Outdoor air DB & WB Return air DB & WB Mixed air entering coils, DB & WB Supply air leaving coils, DB & WB Hot deck Cold deck Air at terminals Conditioned areas DB & WB

Temperatures, Water   Boiler supply and return
                                   Chiller supply and return
                                   Condenser supply and return
                                   Heat exchanger supply and return
                                   Coil supply and return

Refrigerant Temperatures   Hot gas line
                                   Suction line

Overall Building Energy
Readings   At gas meter with all heating on
                                   At electric meter with only lights on
                                   At electric meter with HVAC units on
                                   At electric meter with refrigeration on

Energy conservation must be approached in a systematic manner rather than considering individual items out of context. Systems do not operate in isolation but depend on and react with other systems. It is important to recognize this interaction of systems as modification to one will cause a reaction in another which may be either beneficial or counter-productive.

TEAM EFFORT

A successful energy audit and retrofit must be a team effort of many people beyond the auditor himself.

The building engineer and maintenance personnel must be involved to give the auditor critical information on the actual.conditions of the HVAC systems, on the maintenance history and problems, information on the performance of the systems, changes in the system from original design and what the operating times are. Occupants can provide information, comfort, operation problems and occupancy times.

The design engineer, whether it be the auditor himself or another party, is critical to the evaluation of the audit, redesign and the projection of costs and savings.

Contractors can input valuable information based on experience, actual installation and testing of systems.

Equipment suppliers can input valuable data on the performance and operation of their equipment as well as on actual experience and installations.

The interaction, communications, shared data, previous experience, problem and solutions of all the parties are absolutely required for a audit and retrofit to produce the needed results. No one person can do the job in a vacuum.

HOW TO CHOOSE AN ENERGY ENGINEER

There are several important characteristics that distinguish an effective and reliable energy engineer from one of less competence and reliability.

Experience and proven successful audits and retrofits where energy savings projections were actually attained, where the retrofit costs and pay backs were in line and where the occupants are comfortable and truly satisfied is at the top of the list.

Owning the required auditing and testing instruments, proficiency at taking readings, a knowledge of testing and balancing is an extremely strong asset of the energy engineer.

Open mindedness and imagination are important attributes. Ability to deal effectively and get along with all parties involved and clear, plain language communications are valuable aspects. The ability to sell and persuade others as to the true value of the audit work and energy savings recommendations is vital. A sound background of knowledge in HVAC, electrical, plumbing, etc. is important.

Table 3-3
Consulting Engineering Costs
For Designing HVAC Retrofit Project and
Preparing Drawings and Specifications

RETROFIT COSTS	PERCENT FEE
$5,000 to 25,000	15%
25,000 to 50,000	12%
5,000 to 1000,000	10%
100,000 to 200,000	8%
300 to 500,000	6%
	RANGE 15% to 6%

Table 3-4
Hours And Costs of Complete Energy Audits
Office Buildings, Stores

Audit includes complete field tests of HVAC equipment, lighting and plumbing, surveys, interviews with building personnel, energy history analysis, load calculations, energy conservation proposals, calculations on energy savings, retrofit costs and paybacks.

Size of Building Sq Ft	Typical Tons of Air Cond	Labor		Costs of Audit		
		Man Hours	Hours Per 1000 Sq Ft	Total	Cost Per Sq Ft	Cost Per Ton
1,000	3	5	5.00	$300	$0.300	$100
2,000	6	8	4.00	480	0.240	80
4,000	12	13	3.25	780	0.195	65
6,000	18	16	2.67	960	0.160	53
8,000	24	19	2.38	1,140	0.143	47
10,000	30	21	2.10	1,260	0.126	42
12,000	36	24	2.00	1,440	0.120	40
14,000	42	27	1.93	1,620	0.116	39
16,000	48	30	1.88	1,800	0.113	37
18,000	54	32	1.78	1,920	0.107	36
20,000	60	34	1.70	2,040	0.102	34
25,000	75	40	1.60	2,400	0.096	32
30,000	90	45	1.50	2,700	0.090	30
40,000	120	56	1.40	3,360	0.084	28
60,000	180	78	1.30	4,680	0.078	26
80,000	240	100	1.25	6,000	0.075	25
100,000	30	120	1.20	7,200	0.072	24
200,000	600	160	0.80	9,600	0.048	16
300,000	900	180	0.60	10,800	0.036	12
400,000	1,200	228	0.57	13,680	0.034	11
500,000	1,500	270	0.54	16,200	0.032	11

(*Continued*)

Table 3-4 (*Continued*)
Hours And Costs of Complete Energy Audits
Office Buildings, Stores

Size of Building Sq Ft	Typical Tons of Air Cond	Man Hours	Hours Per 1000 Sq Ft	Total	Cost Per Sq Ft	Cost Per Ton
		Labor		Costs of Audit		
600,000	1,800	312	0.52	18,720	0.031	10
700,000	2,100	350	0.50	21,000	0.030	10
800,000	2,400	368	0.46	22,080	0.028	9
900,000	2,700	387	0.43	23,220	0.026	9
1,000,000	3,000	400	0.40	24,000	0 024	8
1,500,000	4,500	510	0.34	30,600	0.020	7
2,000,000	6,000	600	0.30	36,000	0.018	6
2,500,000	7,500	700	0.28	42,000	0.017	6
3,000,000	9,000	750	0.25	45,000	0.015	5

Correction Factors	Multiplier	Correction Factors	Multiplier
1. Hospitals.	1.30	4. Restaurants.	1.20
2. Schools.	1.20	5. Motels and hotels.	0.90
3. Discount stores.	0.60	6. Industrial plants.	1.10

COSTS based on $40 per hour for field tests by auditing technician and $60 per hour for energy engineer.

Chapter 4

Sample Audits and Forms

The typical 90,000-sq.-ft. suburban office building used as an example in this chapter was built around 1967, before the energy crisis occurred and it incorporates in its design and operation the great energy waste of that era. Overall, the HVAC, electrical and plumbing systems in the building consumed 276,000 Btu per sq ft per year for a total consumption of 24.9 bill Btu for 1991. The energy costs were $3.80 per sq ft and totaled $339,800 for the year.

The targeted energy reduction is 50 percent, reducing the Btu per sq ft from 276,000 down to 138,000 Btu per sq ft with savings of about $1.90 per sq ft or $171,000 per year.

PROBLEMS WITH EXISTING HVAC SYSTEMS AND BUILDING

1. The building has two energy wasting HVAC systems which simultaneously heat and cool, an interior high pressure terminal reheat system and a high pressure perimeter induction system with about one third primary air taken from the outside.

2. The 310 ton chiller, 6,000,000 Btuh boiler and the hot and chilled water pumps must run the year round because of the terminal reheat system, wasting a great deal of energy.

3. The computer room operates 24 hours a day weekdays with a constant cooling load demand forcing the chiller to run the year round and sporadically at nights.

4. The chiller and boiler are oversized; they cycle and run at inefficient levels.

5. Excessive minimum outside air is brought into the building. The settings of the dampers are off and they leak.

6. The lighting levels are excessive.

7. Thermostats are set too high or low in winter and too low in summer.

8. The oil fired boiler is inefficient, scaly and has a poor combustion efficiency, of 60 percent.

9. Maintenance is poor. Filters, coils and strainers are generally dirty. Many control valves, automatic dampers and thermostats are out of calibration, malfunctioning or mis-set.

10. Systems are out of balance.

11. Fans are oversized for load, pumping out more air than required and running 24 hours per day year around.

12. Pumps are oversized for load, pumping out more GPM's than required, running on demand 24 hours a day year around.

13. Starting and stopping times of HVAC equipment not optimized.

14. Paying higher demand rates than need be.

15. Power factors on underload motors not controlled.

Figure 4-1

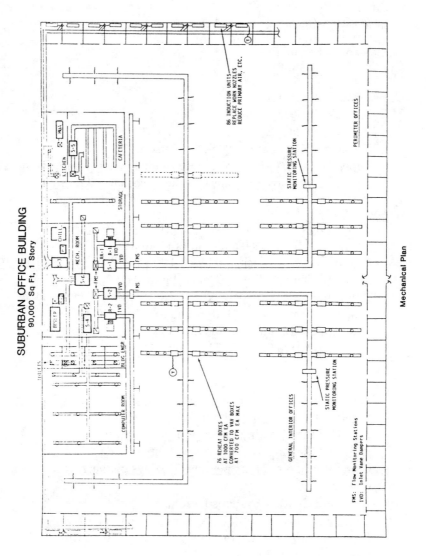

SUBURBAN OFFICE BUILDING
90,000 Sq Ft, 1 Story

Mechanical Plan

Sample 4-1
Building And System Description

Name: Suburban Office Building
Location:
Latitude: 41°N Elevation: 658 Ft When Built 1967

A. Category of Structure
   Office Building

B. Building Description:
   Area, Sq Ft: 90,000 Number of Floors: 1
   Volume, Cu Ft: 1,260,000
   Number of Occupants: 400 Sq Ft/Person: 225
   Types of Areas: Offices, computer room, kitchen, dining room,
      employee lounges, storage, mechanical room

C. Construction Details:
   Glass: Single pane clear U=1.13. no shading, no drapes or
      Blinds, sealed aluminum frame
   Exterior Walls: 8" brick and block. lathe and Plaster. R-6, factor.167
   Roof and Ceilings: Built up tar and gravel on 2" rigid insulation
   (R-6 & metal deck suspended acoustical ceiling. U factor.11
   Floors: Concrete slab, 2 Btuh per sq ft
   Total Exposed Wall Area Sq Ft: 10,430
   Total Glass Area Sq Ft: 6,650 Percent: 39

D. Hours of Occupancy and Operation
   Working Hours: 8 am to 6 pm weekdays 9 am to noon Sat.
   Lighting Hours: 8 am to 9 pm weekdays, 3 hrs Sat.
   HVAC Hours: Cooling and heating year around
   Janitorial Cleanup Times: 6 pm to 9 pm weekdays
   Computer Room: 24 hrs per day, 365 days per year
   Other:

E. Heating and Cooling Systems Description
   Interior office spaces, high pressure terminal reheat system
   High pressure perimeter induction system, 100% primary air
   Chilled water cooling: 1 centrifugal chiller
   Hot water heating oil fired
   Kitchen Dining: Single zone low pressure
   Computer room: Single zone low pressure
   Other:

F. Annual Energy Consumption

Total Heating,Cooling, Electrical, Lighting Per Yr:

Total Btu: 24.9 Bill.	Btu Per Sq Ft: 276,600
Total Energy Cost $454,210	Cost Per Sq Ft: $5.05
Electrical, Total kWh: 3,829,411 kWh/Sq Ft: 42.55	
Total Elec. Costs $383,084	Costs/Sq Ft.: $4.26
Heating Fuels, Btu Per Yr: 11.8 Bill	Per Sq Ft.: 131.110
Total Fuel Costs: $71,800	Costs/Sq Ft: $.81

G. Original Environmental Design Conditions

Heating

Peak Heat Loss Btuh: 4. 8 Mill. Output Degree Days: 6, 000

Design Temperatures:-10°F, 74°F

Avg Winter Temp.: 35 Avg Winter Hours: 4, 800

Cooling

Peak Heat Gain Btuh: 310 tons Degree Days: 682

Design Temperatures: 94°F DBT, 75°F WBT Outdoors
        74°F DBT 50% RH Indoors

Avg Summer Temp.: 75 Avg Summer Cool Hours: 900

Air and Hydronic Flows

Supply Cfm: 121,000 Cfm/Sq Ft: 1. 34

Exhaust Air Cfm: 7,300 Exh Air/Sq Ft..08

Min Outside Air Cfm: 30,000 OA Per Person,Sq Ft: 75 or. 3/sq ft

Make Up Air Cfm: 5,400

HVAC GPM: 1,954   Domestic GPM:_____

H. Lighting

Levels in Foot Candles: 100-200

Levels in Watts/Sq Ft: 4. 0 avg.

Type: Fluorescent

I. Electrical Service

Type: Underground Metering: Primary

Voltage: 277/480V, 3 phase, 4 wire, wye

J. Connected Electrical Loads (kW's)

Lighting: 360 kW Office Equipment: 37+10 comp kW

Heating and Cooling Equipment: 286 kW

Air Handling and Exhausts: 160 kW

Cooking: 50 kW Machinery:_____

Total: 903 kW

HVAC EQUIPMENT SCHEDULE

Job: Corp. Office Bldg.   Location:

SYSTEM	SERVES	EQPT LOCAT	TYPE EQPT	☒ FAN ☐ CFM ☐ GPM	RPM	☒ SP ☐ FT HD	OUTLET VELOC	MOTOR HP	AMPS	VOLTS	COOLING COIL MBH	TONS	GPM	HEATING COIL MBH	GPM
S-1	Interior Office East		AHU	38,000	1585	8"		75	96	460	1140	95	228	800	80
S-2	Interior Office West		AHU	38,000	1585	8"		75	96	460	1140	95	228	800	80
S-3	Perimeter Offices		AHU	7,000	2832	7"		10	28	230	600	50	120	760	76
S-4	Computer Rm.		AHU	10,400		2"		10	28	230	312	26	62	125	13
S-5	Cafeteria & Kitchen		AHU	9,800		1½"		5	15	230	300	25	48	70	70
S-6	Toils. Stor. Mech. Rm.		MZ	7,200		1½"		5	15	230	216	18	43	89	89
	TOTALS			121,200				180			3708	303	729	2644	408
MUA-1	Kitchen		MUA	5,400	760	1"		3	9	230	---	---	---	467	---
	MOTORS:		3 P.H.,	1 Cycle,	1750 RPM										

FAN EQUIPMENT SCHEDULE

Job: Corp. Office Bldg   Location:

SYSTEM	SERVES	EQPT LOCAT	TYPE EQPT FANS	☒ FAN ☒ CFM ☐ GPM	RPM	☒ SP ☐ FT HD	OUTLET VELOC	MOTOR HP	AMPS	VOLTS	COOLING COIL MBH	TONS	GPM	HEATING COIL MBH	GPM
R-1	Interior Office East		Centrif	32,000	509	2"		15	42	230					
R-2	Interior Office West		Centrif	32,000	509	2"		15	42	230					
TE	Toilets		Pre	675	700	3/8"		1/4	3	115					
KE	Kitchen Hood		Pre	5,400	760	1"		3	9	230					
E-1	Conference Rm		Wall	600	960	1/4"		1/4	3	115					
E-2	Conference Rm		Wall	600	960	1/4"		1/4	3	115					
	TOTALS			71,275				34							

PUMP EQUIPMENT SCHEDULE

Job: Corp. Office Bldg   Location:

SYSTEM	SERVES	EQPT LOCAT	TYPE EQPT	☐ FAN ☐ CFM ☒ GPM	RPM	☐ SP ☒ FT HD	OUTLET VELOC	MOTOR HP	AMPS	VOLTS	COOLING COIL MBH	TONS	GPM	HEATING COIL MBH	GPM
P-1	Chilled Water		Pump	744	1750	93		30	40	460					
P-2	Hot Water		Pump	466	1750	80		20	27	460					
P-3	Cooling Tower		Pump	744	1750	93		30	40	460					
CH-1	Chiller							300 (224w)	339	460		310	744		

PROPOSED ENERGY SAVINGS RENOVATIONS

1. Perform a thorough detailed audit with complete test readings, engineering calculations and pricing, after which decisions are made on what retrofit work will, be estimated and performed.

2. The reheat system will be converted to variable air volume (VAV). Some reheat coils will be eliminated. An energy management system will be installed to automatically control fan volume etc.

3. The induction system fan will be turned off in winter and run at a reduced rate only during occupancy during summer.

4. Lighting levels will be reduced from 4 watts per sq ft to 3.

5. Winter space thermostat settings to be reset from 74°F to 70°F and summer from 74°F to 78°F. All stats to be checked and recalibrated if needed.

6. Fan Cfm's will be reduced by reducing RPM's.

7. Pump GPM's will be reduced by replacing pump.

8. Minimum outside ventilation air will be reduced from 30,000 Cfm to 13,000 Cfm.

9. All filters, coils, strainers, condenser tubing and boiler tubing to be checked and cleaned as required.

10. Change computer room unit from a CHW to DX system and reduce the tonnage from 26 to 18 tons. Also add a economizer section on to the unit and include staging in the condenser.

11. The boiler will be cleaned, checked and tuned up.

12. Install computer controlled optimizer on chiller.

13. Install automatic combustion controls and temperature reset controls on boiler.

14. An energy management system will be installed to control on and off times of lighting, fans, HVAC units, pumps, chillers and to minimize demand costs.

15. The systems will be rebalanced, air and water, for maximum efficiency of operation and optimum minimum energy consumption.

16. The systems will then be monitored to keep track of consumption and adjustments and modifications made as required.

Table 4-1
Electrical Consumption History

Building   Suburban Office Building   Year   1992
                      Size Sq Ft   90,000

Electrical Costs

Month	No Of Days	kWh Used	Cost Per kWh	Demand Peak	Demand Charge	Power Factor Adj	Fuel Adj	Total Cost
Jan		262,651	0.0957		$1,099		0.200	$25,136
Feb		285,739	0.0857		941		0.200	$24,491
Mar		219,792	0.1031		1,099		0.417	$22,669
Apr		230,782	0 0957		941		0 417	$22,086
May		311,006	0.1086		2,041		0.200	$33,775
June		362,657	0 1114		2,041		0 200	$40,400
July		403,318	0.1028		1,884		0.200	$41,461
Aug		429,693	0.1028		2,041		0.200	$44,172
Sept		422,001	0.1071		2,198		0.200	$45,196
Oct		350,568	0.0986		1,884		0.200	$34,566
Nov		275,839	0.0886		1,099		0.200	$24,439
Dec		270,161	0.0914		941		0.200	$24,693
Total		3,824,207	0.1002	0	18,209			$383,084
Avg/Mo		318,684	0.1002	0	1,517			$31,924

1.   Average kWh per sq ft of building per year   42.49
     Average electrical cost per sq ft of building per year $4.26
2.   Btuh equivalent for year, (kWh = 3413 Btuh)   13.05 billion
     Btuh average per month   1.09 billion
3.   Average cost per million Btu $29.35
     Average Btu per sq ft of building per year for electrical   13,052
4.   Average kWh per hour   437

Figure 4-3

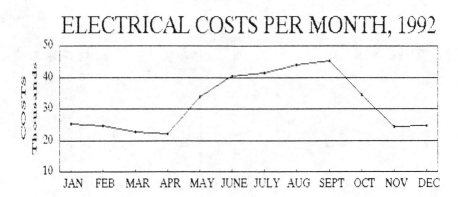

ELECTRICAL COSTS PER MONTH, 1992

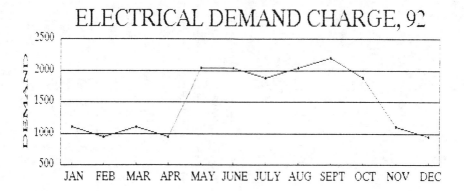

ELECTRICAL DEMAND CHARGE, 92

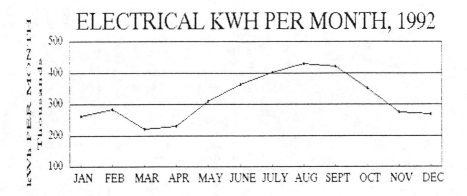

ELECTRICAL KWH PER MONTH, 1992

Table 4-2. Electrical Loads

Building	Suburban Office Building	Existing	New	Date	09-27-1992					
Air Handling Equipment			Seasonal	Summer	Winter		Year Around			
Item	HP Or kWh	Rated		Hours Of Operation Per Year	Actual Average		kWh Per Year	Costs Per Year		
		Volts	Amps	PH		Volt Load	AMP Load			
S-1, AHU, Int. East	75	460	96	3	5,260	460	77	322,315	$32,231		
S-2, AHU, Int. West.	75	460	96	3	5,260	460	77	322,315	$32,231		
S-3, AHU, Perimeter	10	460	28	3	5,260	460	25	104,648	$10,465		
S-4, AHU, Computer RM	10	460	28	3	8,760	460	25	174,280	$17,428		

S-5, AHU, Cafeteria	5	460	5	3	2,500	460	4	7,958	$796
S-6, MZU,Toilets, Storage	5	460	5	3	8,760	460	4 2	29,279	$2,928
MUA-1, Kitchen	3	230	9	3	2,500	230	7	6,963	$696
R-1, Centrifugal Fan	15	230	42	3	5,260	230	35	73,253	$7,325
R-2, Centrifugal Fan	15	230	42	3	5,260	230	34	71,160	$7,116
PRE, Toilet Exhaust	0.25	115	3	1	8,760	115	2	3,486	$349
PRE, Kitchen Exhaust	3	230	9	3	2,600	230	8	8,276	$828
E-1, Wall Fan Conf.	0.25	115	3	1	2,600	115	2.5	1,293	$129
E-2, Wall Fan, Conf.	0.25	115	3	1	2,600	115	2.5	1,293	$129

(S-1, S-2, S-3, R-1, R-2 off nights and weekends, April thru Nov.)

Total	217	369					303.2	1,126,520	$112,652

Costs Per kWh = $0.100

$$kWh/YEAR = \frac{1.73 \times I \times E \times Hours}{1000}$$

Table 4-3
Electrical Loads

Building	Existing Suburban Office Building					New		Date 09-27-1992	
Heat, Cooling Equipment and Pumps			Seasonal		Summer	Winter		Year Around	
Item	HP Or kWh	Rated			Hours Of Operation Per Year	Actual Average		kWh Per Year	Costs Per Year
		Volts	Amps	PH		Volt Load	AMP Load		
P-1, Pump, CHW	30	460	40	3	3,180	460	32	80,981	$8,098
P-2, Pump, HW	20	460	27	3	3,180	460	22	55,674	$5,567
P-3, Pump, Condenser	30	460	40	3	3,180	460	32	80,981	$8,098
Chiller	310	460	339	3	3,180	460	290	733,887	$73,389
Cooling Towers	9	230	28.8	3	3,180	230	23	29,102	$2,910
Total	399		474.8				399	980,625	$98,062

Costs per kWh = $0.10

$$kWh/Year = \frac{1.73 \times I \times E \times Hours}{1000}$$

Table 4-4
Electrical Loads Recap

Building Suburban Office Building

Date 11-18-1992

Item	Connected Load kWh	HP	PH	Existing / New (Year Around)	Hours Of Operation Per Year	Summer Actual Load kWh	BHP	Winter Actual kWh Per Year	Costs Per Year	Percent Of Total Peak
Lighting, Winter	360				2,200	360		792,000	$79,200	0.21
Summer	360				1,800	360		648,000	$64,800	0.17
Air Handling Eqpt.		216						1,126,520	$112,652	0.29
Chillers, Pumps, Cool.Twrs		384						980,625	$98,063	0.26
Computers	10				8,760			87,600	$8,760	0.02
Office Equipment	37	25			2,600			65,000	$6,500	0.02
Kitchen Equipment	50				2,500			125,000	$12,500	0.03
Total	903	625						3,824,745	$382,475	1.00

3,829,411 kWh/90,000 sq ft = 42.55 kW/sq ft/yr

$382,475/90,000 sq ft = $4.25/sq ft/yr

3,829,411 kWh × 3416 Btu/kW = 13.1 Billion equivalent Btu

Costs Per kWh = $0.100

1 HP = 746 Watts

Sample 4-2. Peak Heating, Cooling and Cfm Per Area

Building   Suburban Office Building   X Existing ____ New   Date 11-18-1992

Total Sq Ft 90,000   Average Cfm Per Ton 400

Area	Sq Ft of Area	Cooling Load		Heating Load		Cfm Supply		Direct Ext.
		Sq Ft Per Tons	Tons	Btu Per Sq Ft	Mbh	Cfm Per Sq Ft	Cfm	Cfm
Interior Offices, East	31,275	330	95	25	782	1.20	37,909	
Interior Offices, West	31,275	330	95	25	782	1.20	37,909	
Perimeter Offices, East	3,489	250	14	60	209	1.60	5,582	
Perimeter Offices, South	4,950	225	22	60	297	1.60	8,800	
Perimeter Offices, East	3,489	250	14	60	209	1.60	5,582	
Conference Room, East	450	250	2	60	27	1.60	720	1,200
Conference Room, West	450	250	2	60	27	1.60	720	1,200
Vestibule	225	100	2	35	8	4.00	900	

Computer Room	3,600	140	26	35	126	2.90	10,286	
Men's Toilets	224	400	1	53	12	1.00	224	225
Women's Toilets	224	400	1	53	12	1.00	224	225
Women's Lounge	224	400	1	53	12	1.00	224	225
Hallway	300	400	1	35	11	1.00	300	
Building Engineer	225	400		35	8	1.00	224	
Storage Room	1,800	650	3	35	63	0.61	1,108	
Mechanical Room	4,800	400	12	35	168	1.00	4,800	
Kitchen	1,000	1,43	7	60	60	2.80	2,797	5,400
Cafeteria	2,000	150	13	35	70	2.70	5,333	
Total	90,000	290	310	31	2,883	1.37	123,642	8,475

Sample 4-3
Heat Loss Calculation Form

[ ]Peak Per hr   [X]Seasonal   [X]Existing   [ ]New

Building   Date   11-18-92
Location   Suburban Office Building   Latitude   41
Type Building   Office and Labs   Stories   1   When Built   1967
Sq Ft Area   90,000   Cubic Ft of Space   1,260,000

_×_ Calculation for Whole Building   For Partial Area
Budget Load: Sq Ft 90,000 × Btu Yr/Sq Ft 78,500 = 7,065 Mill Btu Yr
Outside Design: DB -10 WB   Avg OA Temp. 35   Winter Hours 4,800
Inside Design: 74 DB(day)   RH   DB(night)

Item	Dimensions	Sq Ft	U	Temp Diff	Btu: Per Hour / Seasonal	Seasonal Hours
Roof or Ceiling	360 × 250	90,000	0.110	39.00	1,853,280,000	4,800
Floor	360 × 250	90,000	0.100	16.00	691,200,000	4,800
Glass	950 × 7	6,650	1.100	39.00	1,369,368,000	4,800
Doors						

Walls	1220×14	10,430	0.330	39.00	644,323,680	4,800
Cold Inside Walls						
Ventilation	1.43 AC/HR Cfm =	30,000	1.080	39.00	6,065,280,000	4,800
Duct Losses						
Gross Total Heat Loss					10,623,451,680	
Heat Gaines, Lights	4W/Sq Ft	90,000	4.000	3.416	3,074,400,000	2,500
People	No. People =		400	250	200,000,00	2,000
Off. Equipment	HP =		25	2,550	127,500,000	2,000
Computers	kW =		5	3416.00	149,620,800	8,760
Total Internal Gaines					3,551,520,800	
Net Total Building Heat Loss					7,071,930,880	78,577*
Input To Heating Equipment, efficiency	=	.60			11,800,000,000	131,110*

Btu/Year = Sq Ft × U × Avg Temp. Diff. × Winter Hours

* Btu/Sq Ft/Year

Btuh = Sq Ft × U × Temp. Diff.

Remarks

Sample 4-4
Cooling Load Calculation Form

[X] Peak per hr [ ] Seasonal [X] Existing [ ] New

Building	Suburban Office Building	Date __11-18-92__
Location	Latitude __41__	Peak Load, hr mo:
Type Building	Office And Labs Stories __1__	When Built __1967__
Sq Ft Area __90,000__	Cubic Ft of Space __1,260,000__	
__×__ Calculation for Whole Building	For Partial Area	
Building Load:	Sq Ft __90,000__ __×__	Sq Ft/Ton 270 = 333 Tons
Outside Design:	DB 95 WB 75 Avg OA Temp. __74__	__35__ Summer Hours __900__
Inside Design:	__DB(day)__ __74__ RH DB(night)	__74__

Item (Orient.)	Dimensions	Sq. Ft.	U or Factor*	Temp Diff	Sensible Btuh	Latent Tons	Btuh
Roof or Ceiling		90,000	0.110	20.00	198,000	16.50	
Glass East		1,500		15.00	22,500	1.88	
(Solar) South		2,160		50.00	108,000	9.00	
West		1,500		15.00	22,500	1.88	

Item				Btuh	Tons	Sq Ft/Ton
Glass South (Conduction)	5,160	1.130	20.00	116,616	9.72	
(Conduction)					0.00	
Walls E,W,N 9,040 (Conduction)		0.167	20.00	30,194	2.52	
(Conduction)	2,880	0.167	50.00	24,048	2.00	
Lighting	90,000	4	3.416	1,229,760	102.48	
People, No. (No.) 400		250	100,000	100,000	8.33	
Off. Equipment Motors (hp) 25		2,550		63,750	5.31	
Computers (kW) 10		3,416		34,160	2.85	
Kitchen					0.00	
Ventil. Air, Sens 1.43 AC/Hr	30,000 Cfm	1.08	20.00	648,000	54.00	
Ventil. Air, Lat. 1.43 AC/Hr	30,000 Cfm	0.010	4,840	1,452,000	0.00	
Total Cooling Load, Sensible				2,597,528	216	
Latent				1,552,000	121	
Grand Total				4,149,528	337	267*

Btuh = Sq Ft $\times$ U $\times$ Temp. Diff.

* U, CLTD, CLF, kW, HP, WD

CLTD = (Temp diff)-(Daily Range + 14)/2

Btu/year = Sq Ft $\times$ U $\times$ Avg Temp. Diff. $\times$ hrs

* Sq Ft/Ton

CHILLER TEST REPORT

Job _____ Job No. _____ Date _____

Location _____ System _____

Equipment Location _____ Serves _____ Tested by: _____

COMPRESSOR DATA		
Manufacturer		
Model/Size		
Type		
Capacity	tons @	GPM
Refrigerant	Pounds	
KW	KW Per Ton	
Serial No.		

COMPRESSOR MOTOR			
Manufacturer	Serial No.		
Frame No.	Type Frame	□T	□U
Svc. Factor:		Rated	Actual
HP, Nameplate			
BHP [$HPnp \times \frac{Aa}{Ar} \times \frac{Ia}{Vr}$]			
Amps, L_1 L_2 L_3			
Voltage, L_1 L_2 L_3			
RPM			
Phase			

COMPRESSOR	Design	Actual
Suction Pressure		
Suction Temp.		
Discharge Press.		
Discharge Temp.		
Oil Temp/Press.		

STARTER		
Manufacturer	Model	
Size	Class	
Overload: Required Size :		
Actual:		

EVAPORATOR	Design	Actual
Refrig. Pressure		
Refrig. Temp.		
Ent. Water Pressure		
Lvg. Water Pressure		
Ent. Water Temp.		
Lvg. Water Temp.		
Flow GPM		

CONDENSER	Design	Actual
Liquid Line Pressure		
Liquid Line Temp.		
Ent. Water Press.		
Lvg. Water Press.		
Ent. Water Temp.		
Lvg. Water Temp.		
Flow GPM		

CONDITIONS	
Refrigerant Level	
Oil Level	
Percent Cylinders Unloaded	
Chilled Wat. Control Setting	
Condenser Wat. Control Setting	
Low Wat. Cutout Temp. Setting	
Low Pressure Cutout Setting	
High Pressure Cutout Setting	

$$\frac{KWH}{Per\ Year} = \frac{Volts \times 1.73 \times \frac{Avg}{Amps} \times \frac{Yearly\ Hours}{of\ Operation} \times PF}{1000}$$

KW's =
Per Year [   ]

Remarks _____

□ Purge Operation Checked

□ Crankcase Heater Checked

_____

_____

PUMP TEST REPORT

Job _____ Job No. _____ Date _____

Location _____ System _____

Equipment Location _____ Serves _____ Tested by: _____

PUMP DATA		
Manufacturer		
Model /Size		
Type Pump		
Impeller Size		
	Rated	Actual
GPM		
Total Ft.Head		
RPM		

MOTOR			
Manufacturer	Serial No.		
Frame No.		Svc. Factor	
		Rated	Actual
HP, Nameplate			
Amps, L_1 L_2 L_3			
Voltage, L_1 L_2 L_3			
RPM			
Phase			

PUMP PRESSURES		
	Design	Actual
Static Hd(Pump Off)		
Discharge		
Suction		
Block Off: (Running, no flow)		
Discharge		
Suction		
Total		
Running:		
Discharge		
Suction		
Total		

STARTER		
Manufacturer	Model	
Size	Class	
Overload: Required Size :		
Actual:		

$$\text{BHP} \left[\text{HPnp} \times \frac{Aa}{Ar} \times \frac{Va}{Vr} \right]$$

$$\frac{\text{KWH}}{\text{Per Year}} = \frac{\text{Volts} \times \sqrt{3}^* \times \text{Avg Amps} \times \text{Yearly Hours of Operation}}{1000} \times \text{PF}$$

$$= \boxed{} \quad {}^*(3 \text{ phase})$$

Remarks _____

_____

_____

_____

_____

_____

FAN TEST REPORT

Job _____ Job No _____ Date _____

Location _____ System _____

Equipment Location _____ Serves _____ Tested By: _____

☐Air Handling Unit ☐Roof Top Unit ☐Furnace ☐Supply Fan ☐Exhaust Fan ☐Pkg Unit

☐LP ☐MP ☐HP ☐Constant Volume ☐VAV

FAN DATA

Manufacturer	
Model Size	
Type Fan	☐Centrigal ☐Roof Exhaust ☐Inline ☐Vane Axial ☐Prop.
Type Wheel	☐Backward Incline ☐Air Foil ☐Forward Curve ☐Paddle Wheel
Wheel: ☐Alignment OK ☐Gap ☐Fastened ☐Clean	
Belts C to C Distance	
Pulleys: Fan Dia. Mot. Dia.	
Motor Movement	
Bearings☐Zerk☐Seal ☐Cut Off Plate OK	

MOTOR

Manufacturer		Serial No.	
Frame No.		Type	☐T ☐U
Svc. Fact.		Rated	Actual
HP, Nameplate			
BHP			
Amps, L_1 L_2 L_3			
Voltage, L_1 L_2 L_3			
RPM			
Phase			

FAN PERFORMANCE

	Design	Actual
Fan CFM		
Outlet CFM Total		
Fan RRM		
Fan S.P.		

STARTER

Manufacturer	Model
Starter Size	Class
Overload: Required Size	
Actual	

CONDITIONS

Vortex Damper Position				
Outside Air Damper Setting				
Return Air Damper Setting				
Filter Conditions				
Coil Conditions				
Temperatures				
OA:	DB	WB	RH	
RA:	DB	WB	RH	
Mixed Air:	DB	WB	RH	
Discharge	DB	WB	RH	
Space:	DB	WB	RH	
Duct Temp. Drop	DB			

STATIC PRESSURE DROPS

	Upstream	Downstream	Total Drop
Filter			
Heat. Coil			
Cool. Coil			
Fan Inlet			
Fan Discharge			
Total Fan S.P.			

Remarks

BTU ENERGY CONSUMPTION PROFILE
BTU'S IN BILLIONS, TOTAL BTU'S 24.9

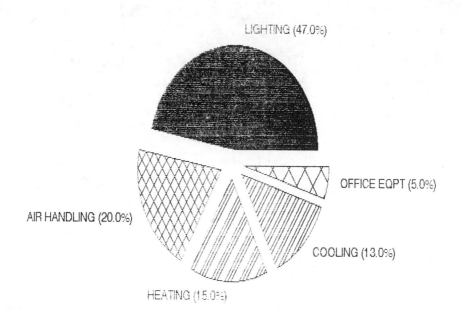

LIGHTING (47.0%)

OFFICE EQPT (5.0%)

AIR HANDLING (20.0%)

COOLING (13.0%)

HEATING (15.0%)

Section 3

Engineering and Financial Evaluations

Chapter 5-
Rule of Thumb Energy Operating Costs

Chapter 6-
Engineering Calculations

Chapter 7-
Financial Evaluations

Chapter 5

Rule of Thumb
Energy Operating Cost

It is very helpful to be able to rough out budget figures on your HVAC energy cost for quick determinations of where you are, where the problems are, to conceptualize the energy conversation process and for thinking clearly about it before being immersed in the heavy details. This chapter gives you rule of thumb figures on budget operating costs of various types of HVAC equipment, what extra resistance in air and hydronic systems costs, plus more. There are also charts with overall budget energy consumption figures for different size buildings, etc.

BUDGET ESTIMATING MOTOR OPERATING COSTS

Problems
- Fans Oversized
- Pumps Oversized
- Excessive Resistance in Air or Water Systems

The extra yearly costs for one additional BHP drawn on a motor based on.10 kWh is:

$900   full time
$450   half time

The extra yearly costs for one additional AMP on a 230V 3 phase motor:

$310   full time
$155   half time

BUDGET ESTIMATING COST
OF EXTRA RESISTANCE IN SYSTEMS

Problems
- Dirty Components
- Old high resistance filters
- Dampers Closed etc.
- Poor ductwork or piping design or installation

The extra yearly costs due to additional resistance in ductwork or piping systems as follows is:

Piping   5 extra feet of head = 1 extra BHP
Ductwork   extra 1/2 inch static pressure = 1 extra BHP

$900/yr per HP   full time at 10¢ kWh
$450/yr per HP   half time at 10¢ kWh

BUDGET ESTIMATING BOILER OPERATING COSTS

Problems
- Operating at 40 to 70 percent combustion efficiencies
- Running at 40 to 70 percent partial load
- Dirty tubes

The extra costs per year running 10 percent under design efficiency based on 4800 hour winter avg 35°F outside temperature, running half the time, is:

	Gas	Oil	Electric
Per Million Btuh Input	$1,200	$1,700	$6,000

BUDGET ESTIMATING CHILLERS OPERATING COSTS

Problems
- Operating at 40 to 70 percent of full loads
- Running at lower efficiencies than design
- Dirty condenser tubes

The extra costs for differences in kW per ton over the design rating, based on 1000 hrs of operation and 8 cents per kWh electrical costs is:

$10/yr   per 0.1 kW/ton

Sample of operating costs per ton per year for various kW/ton ratings is:

		Loss
@.7 kW/ton	$70/yr/ton	$0/yr/ton
@.8 kW/ton	$80/yr/ton	$10/yr/ton
@ 1.0 kW/ton	$100/yr/ton	$30/yr/ton
@ 1.2 kW/ton	$118/yr/ton	$48/yr/ton

Hence, a 500 ton chiller at 1.2 kW/ton can lose $24,000 per year over a .7 kW/ton unit.

BUDGET ESTIMATING
AIR COOLED DX CONDENSERS OPERATING COSTS

The electrical costs for running an air cooling condenser unit with a SEER rating of 8 during the summer at $.10 per kWh is:

1000 Cfm............about $125 per ton

BUDGET ESTIMATING LIGHTING ENERGY COSTS

General Problems With Lighting
- Ten to fifty percent too much lighting provided in various areas.
- Lighting not always turned off during unoccupied times.
- Inefficient lighting.

Yearly Costs And Savings
Yearly savings reducing from 4W to 3W per sq ft of building at .08¢/kWh and 4400 hr/yr operation.

@ 4W/sq ft	$1.41 sq ft/yr
@ 3W/sq ft	$1.06 sq ft/yr

$0.35 sq ft/yr savings

Hence, a 90,000 sq ft building reduced from 4 kW to 3 kW per sq ft of building would save 90,000 sq ft × $.35 = $31,500.

BUDGET ESTIMATING OUTSIDE AIR ENERGY COSTS

Problems

Drawing in more minimum outside air than needed through the outside air intake or through infiltration when heating or cooling equipment is running.

The extra costs per year to heat 1000 Cfm of excessive minimum outside air based on 4800 hour winter and avg 35°F outside air temperature is:

Full Time In Winter per 1000–Cfm (2400 Hr, .75 Eff.)

Gas	$1,200/yr
Oil	$1,700/yr
Electric	$4,500/yr

Half Time In Winter Per 1000 Cfm (2400 Hr, .75 eff.)

Gas	$600/yr
Oil	$850/yr
Electric	$2,250

Full Time, Summer 1000 Hr, 75°F Av Temp, Per 100 Cfm

Electric Cooling	$1,810

Table 5-1
Btu Heating Consumption And Costs Per Year For Different Size Buildings And energy Consumption Rates

Building Size	Yearly Energy Consumption, Mill Btu's			Gas Costs at $5.00 Per Million Btu
	Btu Per Sq Ft Per Year			
Sq Ft	25,000	50,000	75,000	
1,000	25	50	75	$250
2,000	50	100	150	500
3,000	75	150	225	750
4,000	100	200	300	1,000
6,000	150	300	450	1,500
8,000	200	400	600	2,000
10,000	250	500	750	2,500
15,000	375	750	1,125	3,750
20,000	500	1,000	1,500	5,000
25,000	625	1,250	1,875	6,250
30,000	750	1,500	2,250	7,500
35,000	875	1,750	2,625	8,750
40,000	1,000	2,000	3,000	10,000
50,000	1,250	2,500	3,750	12,500
70,000	1,750	3,500	5,250	17 500
90,000	2,250	4,500	6,750	22,500
100,000	2,500	5,000	7,500	25,000
150,000	3,750	7,500	11,250	37,500
200,000	5,000	10,000	15,000	50,000
300,000	7,500	15,000	22,500	75,000
400,000	10,000	20,000	30,000	100,000
500,000	12,500	25,000	37,500	125,000
600,000	15,000	30,000	45,000	150,000
700,000	17,500	35,000	52,500	175,000
800,000	20,000	40,000	60,000	200,000
1,000,000	25,000	50,000	75,000	250,000
2,000,000	50,000	100,000	150,000	500,000
3,000,000	75,000	150,00	225,000	750,000

Actual Heat Loss takes into consideration all heating losses and gains during winter.

Table 5-2
Yearly Heating Costs Per Sq Ft of Roofs, Walls, Etc. For Different R Factors And Fuel Costs

Avg. Transmission Factor of Roof, Wall		Heating Cost per sq. ft. of Roofs, Walls, etc. Fuel Cost per Million Btu						
R	U	$4	$6	$8	$10	$15	$20	$25
0.88	1.130	$1.012	$1.514	$2.023	$2.531	$3.797	$5.062	$6.328
1.00	1.000	0.896	1.340	1.790	2.240	3.360	4.480	5.600
1.80	0.550	0.493	0.737	0.985	1.232	1.848	2.464	3.080
2.00	0.500	0.448	0.670	0.895	1.120	1.680	2.240	2.800
3.00	0.330	0.296	0.442	0.591	0.739	1.109	1.478	1.848
4.00	0.250	0.224	0.335	0.448	0.560	0.840	1.120	1.400
5.00	0.200	0.179	0.268	0.358	0.448	0.672	0.896	1.120
6.00	0.160	0.143	0.214	0.286	0.358	0.538	0.717	0.896
7.00	0.140	0.125	0.188	0.251	0.314	0.470	0.627	0.784
8.00	0.125	0.112	0.168	0.224	0.280	0.420	0.560	0.700
9.00	0.111	0.099	0.149	0.199	0.249	0.373	0.497	0.622
10.00	0.100	0.090	0.134	0.179	0.224	0.336	0.448	0.560
12.00	0.083	0.074	0.111	0.149	0.186	0.279	0.372	0.465
14.00	0.071	0.064	0.095	0.127	0.159	0.239	0.318	0.398
16.00	0.062	0.056	0.083	0.111	0.139	0.208	0.278	0.347
18.00	0.055	0.049	0.074	0.098	0.123	0.185	0.246	0.308
20.00	0.050	0.045	0.067	0.090	0.112	0.168	0.224	0.280

22.00	0.045	0.040	0.060	0.081	0.101	0.151	0.202	0.252
24.00	0.041	0.037	0.055	0.073	0.092	0.138	0.184	0.230
26.00	0.038	0.034	0.051	0.068	0.085	0.128	0.170	0.213
28.00	0.035	0.031	0.047	0.063	0.078	0.118	0.157	0.196
30.00	0.033	0.030	0.044	0.059	0.074	0.111	0.148	0.185
35.00	0.028	0.025	0.038	0.050	0.063	0.094	0.125	0.157
40.00	0.025	0.022	0.034	0.045	0.056	0.084	0.112	0.140
45.00	0.022	0.020	0.029	0.039	0.049	0.074	0.099	0.123
50.00	0.020	0.018	0.027	0.036	0.045	0.067	0.090	0.112

Basis

1. Based on: 4,800 hours winter seasons in Northern Half of the U.S., inside temperature of 70°F, average outside temperature of 35°F.
2. Based on Btu input to heating equipment with any efficiency of 75 percent.

Heating Pickups Not Compensated For:

1. Motor operating heat pickup not included.
2. Pickup from lighting heat not included. To include multiply by .75.
3. People heating pickup not included.

Other Building Loads

1. To include average building electrical costs for HVAC equipment, motors and lighting per sq. ft. per year multiply the above per sq.ft. cost by 1.6. This covers all motors and lighting.

Sample Calculation 5-1
Winter Heat Loss And Operating Costs Per Sq. Ft.
Of Building Skin For different U Factors

With Average .30 U Factors (R 2.5)

$$\text{BTU/Sq. Ft./yr} = \frac{35TD \times 4800HR \times .30}{.75} = 67,200$$

Cost/Sq. Ft./YR = 67,200 × $5/Mill Btu = $.34

With Average .20 U Factor (R 5)

$$\text{BTU/Sq. Ft./yr} = \frac{35TD \times 4800HR \times .20}{.75} = 44,800$$

Cost/Sq. Ft./Yr = 44,800 × $5/Mill Btu = $.224

With Average .15 U Factor (R 7)

$$\text{BTU/Sq. Ft./yr} = \frac{35TD \times 4800HR \times .15}{.75 \text{ eff}} = 33,600$$

Cost/Sq. Ft./Yr. = 33,600 × $5/Mill Btu = $.17

With Average .10 U Factor (R 10)

$$\text{BTU/Sq. Ft./yr} = \frac{35TD \times 4800HR \times .10}{.75 \text{ eff}} = 22,400$$

Cost/Sq. Ft./Yr. = 22,400 × $5/Mill Btu = $.112

Based on Chicago area winter:
    Average outdoor temperature: 37°F
    Indoor design temperature: 72°F
    Average temperature difference (TD): 35°F
    Length of winter: 200 days × 24 hours = 4800 hours
    Combustion efficiency: .75
    Cost per million Btu: $5.00

Where:
    U = Btu/Sq. Ft./°F
    TD = Temperature Difference
    HR = Hours

Chapter 6

Engineering Calculations

Seasonal heating and cooling energy consumption calculations are different than the normal peak load heating and cooling load calculations.

The energy consumption of a building is based on the amount of energy used over a period of time, generally a season or a year. However, the peak heating, cooling load for a building is based on the maximum amount heating and cooling needed at the peak loads. The premise of the peak load is to determine the worst heating and cooling loads generally encountered during the year and to provide a system which has sufficient capacity to meet this maximum design load. It says nothing about the total energy consumption for the year.

This chapter covers formula's and charts on seasonal energy consumption, peak load calculations, internal heat gains, R factors, weather data, costs, degree days, bin methods, cooling hour loads, total energy savings, etc.

CALCULATING BUILDING ENERGY CONSUMPTION

Rate of Seasonal Heat Loss Or Gain (Btu Per Sq. Ft. Per Year)

The range for the rate of seasonal heat loss for winter of a building can vary astronomically. Office buildings in the U. S. may be as low as 20, 000 Btu/sq. ft./yr. and as high as 150, 000 Btu/sq. ft./yr. and may average 30, 000.

Hence, a 90, 000 sq ft. building may have a total seasonal heat loss of 1. 8 billion Btu in a well insulated building in the southern part of the U.S., while it may be six times as much, 10.8 billion Btu in a leaky poorly insulated single story building in the north.

Table 6-1
Average Annual Energy Performance
In Btu's Per Square Feet

Building Type	National	1	Heating and Cooling Degree Day Region*					
			2	3	4	5	6	7
Office	84,000	85,000	76,000	65,000	61,000	51,000	50,000	64,000
Elementary	85,000	114,000	70,000	68,000	70,000	53,000	48,000	57,000
Secondary	52,000	77,000	65,000	55,000	51,000	37,000	41,000	34,000
College/Univ.	65,000	67,000	70,000	46,000	59,000			83,000
Hospital	190,000		209,000	171,000	227,000	207,000		197,000
Clinic	69,000	84,000	72,000	71,000	65,000	61,000	59,000	59,000
Assembly	61,000	58,000	76,000	68,000	51,000	44,000	68,000	57,000
Restaurant	159,000	162,000	178,000	186,000	144,000	123,000	137,000	137,000
Mercantile	84,000	99,000	98,000	86,000	81,000	67,000	83,000	80,000
Warehouse	65,000	75,000	82,000	65,000	50,000	38,000	37,000	39,000
Residential Non-Housekeeping	95,000	99,000	84,000	94,000	125,000	90,000	93,000	106,000
High Rise Apt.	49,000	53,000	53,000	52,000	53,000	84,000	20,000	

*See Auditing Procedures, Chapter 3, pg. 3-3

Total Seasonal

The total seasonal energy consumption is different than the peak hourly energy consumption which is figured to cover the worst possible condition that may be encountered in the winter or summer. The seasonal energy consumption deals with averages of temperatures, heat transmission factors, degree days and time spans. It also takes into considerations the heat gains generated in the building by lights, people etc. which is not a factor in the peak heat loss or gain.

THE ENERGY CONSUMPTION
OF A BUILDING IS DEPENDENT ON:

1. R values of walls, windows, roofs, doors and resultant "U" factors.

2. Percentage of glass and R factors.

3. Amount of outside air being brought in either by the supply fan or by infiltration or both.

4. Amount of air being exhausted.

5. Size of building.

6. Ratio of building skin surface to floor sq footage.

7. Solar orientation.

8. Average outside and indoor temperature. Average outdoor and indoor humidity levels.

9. Lighting levels and operational times.

10. Occupancy levels and operational times.

11. Efficiency of equipment and degree of maintenance.

12. Internal heat gain due to lighting, occupancy, equipment and solar heat.

RATIO OF SKIN-TO-FLOOR AREA FACTOR

The more outside skin surface there is in relationship to sq ft of floor area, such as with a single story building, the greater the rate of energy consumption. The smaller the ratio, as with multi-story buildings, the lesser the rate of energy consumption.

For example:

A single story building with a ratio of 1.5 skin surface to floor surface area. Rate of heat losses or gains may be 50 to 100% greater than multi-story buildings.

A two story building with a 1 to 1 ratio. Energy consumption may be 25% less than single story.

Multi-story buildings with .5 to one ratio. Rate of energy consumption may be 50% less than single story.

OVERALL BUILDING SKIN FACTOR

The yearly skin loss formula is: (does not include OA and infiltration losses nor building heat gains).

$$\text{Btu/Yr} = \frac{\text{(Degree)}}{\text{Days}} \times \text{(Hours/day)} \times \frac{\text{(U Factor)}}{\text{Average}} \times \text{(Sq Ft Skin)}$$

Example for 90,000 sq ft, single story. building, 14 ft high.

Btu/Yr　　= 6600 × 24 ×.2 average U × 107,080 Sq ft
　　　　　= 3.393 Bill Btu (avg 37,692 Btu/sq ft/yr)

WINTER INTERNAL HEAT GAINS

The annual heat loss of a building structure is reduced by internal heat gains in the winter as follows:

1. Lighting heat pickup which may constitute a 25 to 50 percent regain. At 3 watts per sq ft.
    3W × 3.416 Btu/W × 90,000 sq ft × 2500 hr = 2.31 Bill Btu/Yr

2. People heat pick up at occupancy rate of 1 person per 225 sq ft.

$$\frac{90{,}000 \text{ sq ft}}{225 \text{ sq ft/person}} = 400 \text{ people}$$

400 People × 250 Btuh × 2000 Hr = 200 Mill Btu/Yr

3. Solar radiation gains in winter of glass, roof, walls.

4. Equipment heat gains:

1 HP = 2550 Btuh
(25 HP) × (2550 Btuh) × (2000 hrs) = 127 Mill Btu/Yr

If the infiltration for a 90,000 office building is one half an air change per day, the seasonal heat loss is calculated as follows:

$$\text{CFM} = \frac{\text{Air Changes/hr} \times \text{cu ft of bldg}}{60}$$

$$= \frac{.5 \times 630{,}000}{60} = 10{,}500 \text{ CFM}$$

Btu/yr = Cfm × 1.08 × TD × hours

Btu/yr = 10,500 × 1.08 × 30 × 4800 hr = 1.633 Bill Btu

QUICK APPROXIMATION
OF INSTANTANEOUS HEAT LOSSES

A quick, approximate method of determining the overall heat loss of a building is to read the gas meter and relate to outside air temperature as follows:

1. Read gas meter or oil gpm meter.

2. Add calculated lighting, people, solar and equipment heat gains if done during occupancy.

3. Subtract inefficiency of heating apparatus.

4.   Relate to concurrent outside temp.

Example: One Hour Gas Meter Reading
        Average Outside Air 35°F:
        90,000 sq ft building
        3 Watts per sq ft
        400 people

1.   Reading: 2000 cu ft of gas in 1 hr = 2 mill Btuh

2.   Calculate heat gains:
     a)   3 watts per sq ft × 90,000 sq ft = 270,000 watts
                                 $270,000 \text{ W} \times 3.413 = 92,150$ Btuh
     b)   Occupancy,   400 people × 250 = 100,000 Btuh
     c)   Take reading in darkness to eliminate calculations of solar gain.
     d)   Building equipment: 45 kWh × 3413 = 153,600 Btuh
                                 Total Heat Gain = 345,750

3.   Divide by efficiency of heating equipment

$$\frac{345,750}{.75} = 461,000 \text{ input}$$

4.   Actual Building Btuh Heat Loss at 35°F OA =   2,000,000
                                                     +461,000
                                             Total   2,461,000

5.   If it's possible to read gas or oil meter at night during no occupancy
     with all lighting and equipment turned off, total vacancy and with no
     solar load, the meter reading will give a true representation of the
     building heat loss if internal heat gains from HVAC equipment is
     disregarded.

Table 6-2
Budget Estimating Peak Heating, Cooling And Cfm Loads
Per Square Foot of Building

Type Building	Cooling Btu Per Sq Ft	Load Sq Ft Per Ton	Heating Load Btu/ft	Supply Cfm Per Sq Ft	Duct WT lbs/Per Sq Ft
Apartments,					
Condominiums	25	480	26	.8	.2
Auditoriums	40	300	40	1.3	.8
Banks	48	250	26	1.6	1.2
Bowling Alleys	40	300	40	1.3	.6
Churches	36	330	36	1.2	.5
Clubhouses	50	240	40	1.6	1.0
Cocktail Lounges	70	170	30	1.3	1.2
Computer Rooms	140	85	20	4.5	1.5
Colleges: Admin.,					
Classrooms	44	270	42	1.5	1.3
Dormitories	—	—	25	—	.3
Gyms, Fieldhouses	—	—	40	.5	.3
Science Bldgs.	54	220	55	1.8	1.4
Court Houses	50	240	35	1.6	1.2
Fire Stations	—	—	35	—	.7
Funeral Homes	30	400	32	.9	.8
Hospitals	44	170	28	1.4	1.5
Hotels	34	350	28	1.1	.5
Housing for Elderly	—	—	28	—	.2
Jails	25	480	30	.8	1.2
Laboratories	60	200	50	2.0	1.5
Libraries	46	260	37	1.5	1.3
Manufacturing Plants	40	300	36	1.3	.4
Medical Centers,					
Clinics	35	340	35	1.1	1.0
Motels	30	400	30	1.0	.5

(Continued)

Table 6-2 (*Continued*)
Budget Estimating Peak Heating, Cooling And Cfm Loads
Per Square Foot of Building

Type Building	Cooling Btu Per Sq Ft	Load Sq Ft Per Ton	Heating Load Btu/ft	Supply Cfm Per Sq Ft	Duct WT lbs/Per Sq Ft
Municipal Bldgs.,					
Town Halls	45	265	36	1.4	1.2
Museums	34	350	40	1.1	.8
Nursing Homes	43	280	34	1.4	.5
Office Bldgs:					
Low Rise	35	340	35	1.2	.5-.8
High Rise	40	300	30	1.3	1.1
Police Stations	42	285	35	1.3	1.0
Post Offices	44	270	40	1.4	1.2
Project Homes	—	—	28	—	.1
Restaurants	80	150	35	2.0	1.0
Residences	25	500	30*	.8	.3
Schools:					
Elementary	—	—	26	.8	.7
Middle, Jr.Highs	36	333	26	1.0	1.1
High Schools	33	360	40	1.1	1.3
Vocational	20	600	40	.9	1.5
Stores:					
Beauty Shops	53	190	30	2.0	1.3
Department	34	350	30	1.1	.7
Discount	34	350	32	1.1	.4
Retail Shops	50	340	32	1.6	.8
Shopping Centers	30	400	30	1.1	.4
Supermarkets	30	400	32	1.0	.4
Theaters	40	300	40	1.3	.8
Warehouses	—	—	20	—	.2

1. Cooling load based on 15 temp. difference, 50 × RH, 400 Cfm per ton
2. Heating load based on 70 temp. difference and is the output Btu. Apply inefficiency factor to heating equipment for input Btu
3. Price includes all HVAC material, labor and subs and overhead and profit *R-11 wall and R-19 ceiling insulation

Sample
Heat Loss Calculation

[ ] Peak per Hr [X] Seasonal [X] Existing [ ] New

Building Suburban Office Building Date 11-18-92

Location Latitude 41

Type Building Office and Labs Stories 1 When Built 1967

Sq Ft Area 90,000 Cubic Ft of Space 1,260,000

X Calculation for Whole Building For Partial Area

Budget Load: Sq Ft 90,000 × Btu Yr/Sq Ft 78,500 = 7,065 Mill Btu yr

Outside Design: DB-10 WB Avg OA Temp. 35 Winter Hours 4,800

Inside Design: DB(day) 74 RH DB(night)

Item	Dimensions	Sq Ft	U	Temp Diff	Btu: Per Hour Seasonal	Seasonal Hours
Roof Or Ceiling	360 × 250	90,000	0.110	39.00	1,853,280,000	4,800
Floor	360 × 250	90,000	0.100	16.00	691,200,000	4,800
Glass	950 × 7	6,650	1.100	39.00	1,369,368,000	4,800
Doors						

(*Continued*)

Sample — Heat Loss Calculation (Continued)

Item	Dimensions	Sq Ft	U	Temp Diff	Btu: Per Hour Seasonal	Seasonal Hours
Walls	1220 × 14	10,430	0.330	39.00	644,323,680	4,800
Cold Inside Walls						
Ventilation	1.43 AC/Hr Cfm	30,000	1.080	39.00	6,065,280,000	4,800
Duct Losses						
Gross Total Heat Loss					10,623,451,680	
Heat Gains: Lights	4W/Sq Ft	90,000	4.000	3.416	3,074,400,000	2,500
People	No. People =	400	250		200,000,000	2,000
Off. Equipment	HP =	25	2,550		127,500,000	2,000
Computers	kW =	5		3416.00	149,620,800	8,760
Total Internal Gains					3,551,520,800	
Net Total Building Heat Loss					7,071,930,880	78,577
Input To Heating Equipment, efficiency=	.60			11,800,000,000	131,110*	

Btu/Year = Sq Ft × U × Avg Temp. Diff. × Winter Hours
*Btu/Sq Ft/Year

Btuh = Sq Ft × U × Temp. Diff.

Remarks

Sample
Cooling Load Calculation

[X] Peak Per Hour   [ ] Seasonal   [X] Existing   [ ] New

Building __Suburban Office Building__	
Location	Date __11-18-92__
Type Building __Office and Labs__	Peak Load, hr, mo:
Sq Ft Area __90,000__	When Built __1967__
x__ Calculation for Whole Building	Cubic Ft of Space __1,260,000__
Building Load: Sq Ft __90,000__	For Partial Area
Outside Design: DB _95_ WB _75_	x ____ Sq Ft/Ton __270__ = __333 Tons__
Inside Design: DB(day) __74__	Avg OA Temp. _35_ Summer Hours __900__
	RH ____ DB (night) __74__

Item	(Orient.)	Dimensions	Sq Ft	U or Factor*	Temp Diff	Sensible Btuh	Tons	Latent Btuh
Roof or Ceiling	90,000		0.110	20.00	198,000	16.50		
Glass	East		1,500	15.00		22,500	1.88	
(Solar)	South		2,160	50.00		108,000	9.00	
	West		1,500	15.00		22,500	1.88	
Glass	South		5,160	1.130	20.00	116,616	9.72	

(Continued)

Sample
Cooling Load Calculation (Continued)

Item	(Orient.)	Dimensions	Sq Ft	U or Factor*	Temp Diff	Sensible Btuh	Tons	Latent Btuh
(Conduction)							0.00	
Walls	E.W.N		9,040	0.167	20.00	30,194	2.52	
(Conduction)			2,880	0.167	50.00	24,048	2.00	
Lighting			90,000	4	3.416	1,229,760	102.48	
People, No.		(No.)	400	250	250	100,000	8.33	100,000
Off. Equipment		Motors (HP)	25	2,550	63,750	5.31		
Computers		(kW)	10	3,416	34,160	2.85		
Kitchen							0.00	
Ventil. Air, Sens. 1.43 AC hr		30,000	Cfm	1.08	20.00	648,000	54.00	
Ventil. Air, Lat. 1.43 AC/hr		30,000	Cfm	0.010	4,840		0.00	1,452,000
Total Cooling Load,					Sensible	2,597,528	216	
					Latent	1,552,000	121	
					CLTD = (Temp diff			
					Grand Total	4,149,528	337	267*

Btuh = Sq Ft × U × Temp. Diff.
* U, CLTD, CLF, kW, HP, WD
CLTD = (Temp diff)-(Daily Range + 14)/2
Btu/Year= Sq Ft × U × Avg Temp. Diff. × hrs

* Sq Ft/Ton

Table 6-3
Weather Data
Average Seasonal Conditions

		Winter		Summer	
CityState	Avg DB	Length Winter Temp	Avg DB In Weeks	Length Summer Temp	In Weeks
Atlanta	GA	41.1	19.8	78.7	30.0
Boston	MA	35.1	31.1	76.0	19.8
Charleston	SC	43.3	14.2	78.7	36.0
Chicago	IL	34.2	30.0	77.0	20.9
Cleveland	OH	34.0	29.4	76.5	21.0
Dallas	TX	42.5	15.1	82.8	34.6
Denver	CO	35.2	29.4	77.9	22.6
Detroit	MI	33.8	30.5	75.8	19.2
Houston	TX	47.0	6.0	80.3	42.0
Indianapolis	IN	35.8	26.7	78.0	23.9
Las Vegas	NV	43.7	15.6	86.8	35.4
Los Angeles	GA	50.2	8.9	72.0	32.6
Miami	FL	49.3	1.6	80.4	50.1
Milwaukee	WI	33.0	30.0	77.0	20.9
Minneapolis	MN	29.3	31.0	76.8	18.8
Nashville	TN	39.3	23.3	79.7	28.4
New Orleans	LA	46.4	9.4	79.8	39.6
New York City	NY	38.0	27.5	76.0	20.0
Philadelphia	PA	38.2	26.0	77.5	23.7
Portland	OR	44.0	30.8	73.5	15.8
Richmond	VA	40.9	20.9	77.8	26.8
Salt Lake City	UT	36.5	30.2	79.0	19.9
Seattle	WA	43.7	37.3	70.9	9.4
St. Louis	MO	36.1	24.2	79.6	26.3
Trenton	NJ	37.5	26.9	77.1	22.9

Table 6-4
Minimum U's or R's For Walls For Heating

Heating Degree Days	Minimum Acceptable U	R
1000	.40	2.5
2000	.30	3. 33
3000	.30	3. 33
4000	.20	5.0
5000	.20	5.0
6000	.15	7.0
7000	.15	7.0
8000	.10	10.0
9000	.10	10.0

Table 6-5
Minimum U's or R's For Roofs For Heating

Heating Degree Days	Minimum Acceptable U	R
1000	. 30	3.3
2000	. 20	5.0
3000	. 20	5.0
4000	.15	7.0
5000	.15	7.0
6000	.10	10.0
7000	.10	10.0
8000	.06	17.0

Table 6-6
Rates of Heat Gain From Occupants of Conditioned Spaces

Degree of Activity	Total Heat Adults Male Typical Application	Total Heat Adjusted Btu/h	Sensible Heat Btu/h	Latent Heat Btu/h	Btu/h
Seated at rest	Theatre, movie	400	350	210	140
Seated very light work writing	Offices, hotels, apts	480	420	230	190
Seated, eating	Restaurant	520	580	255	325
Seated, light work typing	Offices, hotels, apts	640	510	255	255
Standing, light work or walking slowly	Retail Store, bank	800	640	315	325
Light bench work	Factory	880	780	345	435
Walking, 3 mph light machine work	Factory	1040	1040	345	695
Bowling	Bowling alley	1200	960	345	615
Moderate dancing	Dance Hall	1360	1280	405	875
Heavy work, heavy machine work, lifting	Factory	1600	1600	565	1035
Heavy work, athletics	Gymnasium	2000	1800	635	1165

Table 6-7
R And U Values of Roofing And Wall Materials

Description	Density lb/cu ft	Thickness	R Per inch thickness	Resistance R for thick. shown	Conductance U for thick. shown
AIR					
Outside air film, (15 mph wind)			—	.17	5.88
Inside air film, (still air)			1.64	.61	
Air space, horiz. flow (50°F mean, 30 TD)		1/4"		.91	1.10
		1/2"		1.23	.81
		3-1/2"		1.25	.80
Ceiling air space vert. flow			1.00	1.00	
MASONRY					
Concrete, med. wt.	140.0	1"	.83	.83	1.20
Concrete, light wt.	40.0	1"	.83	.83	1.20
Common brick	120.0	4"	.20	.79	1.27
	120.0	8"	.20	1.59	.63
Face brick	125.0	4"	.11	.43	2.33
Concrete block, L.W.	38.0	4"	.38	1.51	.66
	38.0	8"	.38	.71	1.41
Concrete block	61.0	4"	.50	2.02	.50
	61.0	8"	.50	4.04	.25
Clay tile	70.0	4"	.25	1.01	.99
	70.0	8"	.25	2.02	.50
INSULATION-Fiberglass	.75	1"	3.14	3.14	.32
	.75	3-1/2"	3.14	11.00	.09

Material	Density	Thickness			
Board	.75	6"	3.14	19.00	.005
	5.7	1"	3.33	3.33	.30
	5.7	2"	3.33	6.68	.15
	5.7	3"	3.33	10.00	.10
Cellular glass		4"	2.9	11.6	.09
Polystyrene		4"	4.5	15.0	.06
Polyurethane		4"	6.25	25.0	.04
WOOD–Hardwoods	45.0	1"	.91	.91	1.10
Fir, Pine	37.0	1"	1.19	1.19	.84
	37.0	2"	1.19	2.39	.42
	37.0	2.5"	1.19	2.98	.34
	37.0	3"	1.19	3.58	.28
	37.0	4"	1.19	4.76	.21
Plywood		5/8"	1.25	.78	1.28
		1"	1.25	1.25	.80
Sheathing		5/8"	2.22	1.39	.72
		1"	2.22	2.22	.45
GLASS–Single Pane			—	1.10	.91
Double Pane, 1" Space			—	.52	1.92
FINISHING MATERIALS, WALL.CEILINGS					
Plaster, Gypsum, Foil	100.0	5/8"	.95	.59	1.69
	100.0	3/4"	.95	.71	
Plaster		3/4"			
Acoustic Tile		3/4"	2.34	1.79	.56
METAL–Metal Deck		20 Ga.	—	.33	
ROOFING–Built up		3/8"	—	—	3.03

Formula 6-1
Seasonal Energy Cost Formulas

Heating Cost Per Yr of Fuel For Transmission Losses

$$\text{Fuel Costs/Yr} = \frac{U \times A \times TD \times HR \times \$MMB}{1,000,000 \times EffC}$$

Heating Costs Per Yr of Fuel For Heating Outside Air

$$\text{Fuel Costs/Yr} = \frac{CFM \times TD \times 1.08 \times HR \times \$MMB}{1,000,000 \times EffC}$$

Where:

U	=	Btu/F/Sq Ft, heat transmission
A	=	Area in sq ft
TD	=	Temperature difference between inside and outside surfaces
Hr	=	Hours per season or year
$MMB	=	Cost per million Btu of fuel
effC	=	Efficiency of combustion

1. 4800 seasonal hours typical for northern state winters.
2. 2600 seasonal hours typical for 5 ten hour days of operation.
3. 35°F typical average winter temperature for northern states and 70°F average inside temperature, hence the temperature difference is 35°F.
4. Based on natural gas fuel at $5 per million Btu which equals 50¢ per therm.

Table 6-8
Fuel Heating Values

Fuel	Heating Value
Coal	
anthracite	13,900 Btu/lb
bituminous	14,000 Btu/lb
sub-bituminous	12,600 Btu/lb
lignite	11,000 Btu/lb
Heavy Fuel Oils and Middle Distillates	
kerosene	134,000 Btu/gallon
No. 2 burner fuel oil	140,000 Btu/gallon
No. 4 heavy fuel oil	144,000 Btu/gallon
No. 5 heavy fuel oil	150,000 Btu/gallon
No. 6 heavy fuel oil 2.7 % sulfur	152,000 Btu/gallon
No. 6 heavy fuel oil 0.3% sulfur	143,800 Btu/gallon
Gas	
natural	1,000 Btu/cu ft
liquefied butane	103,300 Btu/gallon
liquefied propane	91,600 Btu/gallon

Source: *Brick and Clay Record*, October 1972, reprinted with permission of the Cahner's Publishing Co. Chicago, IL.

Formula 6-2
Air Heat Transfer Formulas

• **Sensible**

  Btuh = Cfm × temp change × 1.08

Rearranged:

$$CFM = \frac{BTUH\,(\,Sensible)}{1.08 \times temp\,change}$$

Rearranged:

$$Temp\,Change = \frac{Btuh\,(Sensible}{Cfm \times 1.08}$$

• **Latent**

  Btuh = 4840 × Cfm × WH

• **Total Latent and Sensible**

  Btuh = 4.5 × Cfm × HD

Where:

Btuh	=	British thermal units per hour
T	=	Temperature, F
Cfm	=	Cubic feet per minute
EFF	=	Efficiency
HD	=	Difference in enthalpy
WD	=	Difference in humidity ratio (lb water/lb dry air)

Formula 6-3
Steam Formulas

• **Btuh**

$$LB/HR = \frac{BTUH}{1000}$$

• **Equivalent Direct Radiation (EDR)**

LB/Hr = EDR ×.24

• **Steam Coil In Duct**

$$LB/HR = \frac{CFM \times 1.08 \times TDa}{1000}$$

• **Steam To Water Converter**

LB/Hr = gpm × TDw ×.49

• **CV of Steam Valve**

$$CV = \frac{Q \times V?}{63.5 \ h}$$

Where:

Q	=	Lb per hr Steam
V	=	Specific Volume of Steam in cu.ft./lb.
h	=	Pressure Difference Across Valve
TDa	=	Air Temperature Difference
TDw	=	Water Temperature Difference
Specific Heat of Steam	=	.489 Btu/(LB) (F)

Formula 6-4
Electrical Formulas

• **Three Phase Alternating Current Motors**

$$\text{kW Actual (Motor Input)} = \frac{1.73 \times I \times E \times PF}{1000}$$

$$\text{BHP (Motor Input)} = \frac{1.73 \times I \times E \times effM \times PF}{746}$$

$$\text{kWh} = \frac{BHP}{effM}$$

• **If AMPS Are Unknown**

$$\text{AMPS (HP Known)} = \frac{BHP \times 746}{1.73 \times E \times effM \times PF}$$

$$\text{AMPS (kWh Known)} = \frac{kWh \times 1000}{1.73 \times E \times PF}$$

$$\text{AMPS (KVA Known)} = \frac{kVA \times 1000}{1.73 \times E}$$

• **Kilo Volt AMPS**

$$\text{kVA} = \frac{1.73 \times I \times E}{1000}$$

• **Motor Electrical Costs For Year**

$$\text{kWh Input Per Year} = \frac{1.73 \times I \times E \times HR}{1000}$$

$$\text{kWh Input Per Year} = \frac{1.73 \times I \times E \times HR}{1000}$$

Where:

I	= Current in amps	kWh	= Kilowatt hours
E	= Voltage	BHP	= Break horsepower
PF	= Power factor	kVA	= Kilovolt amps
effM	= Efficiency of motor	Hr	= Hour per year

Formula 6-5
Formula For Percent of Accumulated Energy Savings

The actual total savings is not the arithmetical sum of all the percentages saved on the individual energy conservation items. It is less than the sum of individual savings and the following computation must be made working with percentages.

1.

$$\frac{\text{Total Percent}}{\text{Savings}} = 100\left[1 - (1 - \text{Sav}_1) \times (1 - \text{Sav}_2) \times (1 - \text{Sav}_3)...\text{Etc}\right]$$

Where: SAV_1 is percent of saving for each item.

2.   Example problem, turn thermostats down and up:

	Now Setting	Savings
Heating	70%F	6%
Cooling	78%F	15%
Night set back	10%F	7%

$$\frac{\text{Total Percent}}{\text{Savings}} = 100\left[1 - (1 - .06) \times (1 - .15) \times (1 - .07)\right]$$

$$= 26\%$$

HUMAN COMFORT NEEDS

In the region between 2 to 6 feet above the floor and at least 2 feet from any wall, the following conditions are comfortable with average clothing.

1.   The dry bulb temperature shall be between 70° and 72° F in winter and between 76°F and 78°F in summer.

2.   The relative humidity (RH) shall be between 20 and 60 percent.

3.   The air motion passing over the skin of occupants shall be greater than 10 fpm but not exceed 45 fpm.

When "proper" clothing is worn, the upper temperature limit can be extended to 80°F and the lower limit can be dropped to 68°F. Operating in the area of 70° to 72°F winter and 76° to 78°F in the summer is very feasible and occupants learn to adjust their clothing etc. to these conditions.

COST PER MILLION BTU

Cost Comparisons For Different Fuels

There is no inefficiency loss in electrical heating. The input to a electrical heating coil is equal to the actual space heat loss. Gas and oil do have combustion efficiency losses, however, and the building heat loss calculations must be increased for combustion losses, in order to make a valid comparison to electrical heating costs.

1. **Gas:** The typical efficiency of natural gas is 75 percent, hence with input gas at $5.00 per million Btu (50¢ per therm):

$$\frac{\text{Real Cost Per}}{\text{Mill BTU Output}} = \frac{1\text{ Mill BTU}}{.75} = \frac{\$5.00}{.75} = \$6.67$$

2. **Oil:** The typical efficiency of oil is 70 percent, hence with input oil at $7.00 per mill Btu (7 gal at 143,000 Btu per gal):

$$\frac{\text{Real Cost Per}}{\text{Mill BTU Output}} = \frac{1\text{ Mill BTU}}{.70} = \frac{\$7.00}{.70} = \$10.00$$

3. **Electricity:** Since there is no inefficiency factor involved with electric heat, the cost of input equals output. Then at 7¢ per kWh and 3416 Btu per kWh:

$$\frac{\text{Cost Per}}{\text{Mill BTU Output}} = \frac{1\text{ Mill BTU}}{3416} \times 10¢ = 293\text{ kW} \times 10¢ = \$29.30$$

Hence the ratio electric heat runs to gas typically is:
Electric to gas: 3 to 1 ratio
Electric to oil: 2 to 1 ratio

HEATING DEGREE DAYS

Heating Degree days are the number of degrees by which the mean temperature for a day falls short of 65. If, for example, the mean temperature on a given day was 45, that day contributed 20 degree days to the heating season.

The 65°F is assumed to be the actual design temperature required based on the assumption that internal heat gains, such as light, people, equipment etc. will actually raise the indoor temperature to 70°F or 72°F.

This approach is primarily only accurate for residential heating load calculations, but the degree days are useful for ascertaining the severity of summers and winters, and for determining unknown actual number of winter days or hours and average temperature.

Chart 6-1

ANNUAL DEGREE-DAY ZONES

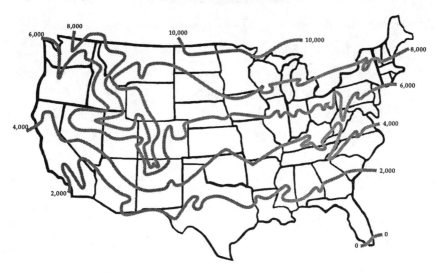

ANNUAL DEGREE-DAY ZONES

To determine the number of unknown winter hours from the given known degree days and average temperature:

If the average outdoor temperature for a particular winter in the Chicago area was 35°F and the given degree days 6000:

Average Temperature Difference = 65 – 35 = 30 TD

$$\frac{\text{Number of}}{\text{Winter Hours}} = \frac{\text{Degree Days}}{\text{Average TD}} = \frac{6000}{30} \times 24 = 4800$$

To determine the average temperature from given know degree days

If number of winter days is 210 and the degree days are 6600:

$$\text{Average Temperature} = \frac{6600}{210} = 32°F$$

To determine the degree days from the given number of winter days and average temperature:

A Midwest city along the 41st parallel may have 210 days of winter that are below 65°F at an average outdoor temperature of 35°F.

Temperature Difference = 65 – 35 = 30

Degree Days = (210 days) × (30 TD) = 6300

Average winter temperature is determined by averaging out the medium temperature between the high and lows per hour for the course of the winter.

$$\frac{\text{Medium Temp}}{\text{Per Hour}} = \frac{\text{High Temp} + \text{Low Temp}}{2} = \frac{34 + 30}{2} = 32$$

Chart 6-2. Annual Occupied Hours At Different Outside Air Temperatures

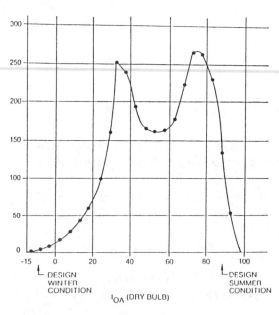

Chicago, Illinois
Bin weather data for a typical year expressed as a curve. As with most locations, peak outdoor conditions are the exception not the rule.

Chart 6-3. Estimated Equivalent Rated Load Hours
For A Normal Cooling Season

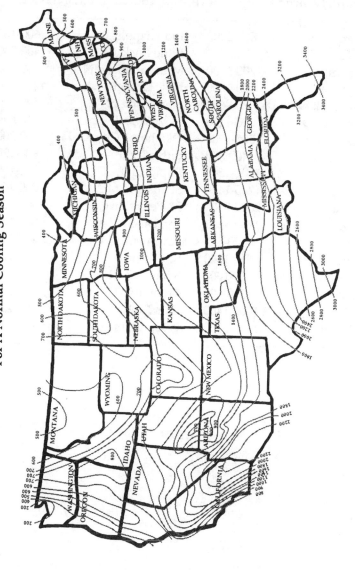

Judge the number of hours according to the line closest to the location of your home.

Reprinted from the June 1976 issue of *ASHRAE Journal* by permission of the American Society of Heating, Refrigerating and Air-Conditioning Engineers, Inc.

Table 6-9
Estimated Equivalent Rated Full Load Hours of Operation For
Properly Sized Equipment During Normal Cooling Season

Albuquerque, NM	800-2200	Indianapolis, IN	600-1600
Atlantic City, NJ	500-800	Little Rock AR	1400-2400
Birmingham, AL	1200-2200	Minneapolis, MN	400-800
Boston, MA	400-1200	New Orleans, LA	1400-2800
Burlington, VT	200-600	New York, NY	500-1000
Charlotte, NC	700-1100	Newark, NJ	400-900
Chicago, IL	500-1000	Oklahoma City, OK	1100-2000
Cleveland, OH	400-800	Pittsburg, PA	900-1200
Cincinnati, OH	1000-1500	Rapid City, SD	800-1000
Columbia, SC	1200-1400	St. Joseph, MO	1000-1600
Corpus Christi, TX	2000-2500	St. Petersburg, FL	1500-2700
Dallas, TX	1200-1600	San Diego, CA	800-1700
Denver, CO	400-800	Savannah, GA	1200-1400
Des Moines, IA	600-1000	Seattle, WA	400-1200
Detroit, MI	700-1000	Syracuse, NY	200-1000
Duluth, MN	300-500	Trenton, NJ	800-1000
El Paso, TX	1000-1400	Tulsa, OK	1500-2200
Honolulu, HI	1500-3500	Washington, DC	700-1200

Table 6-10
Cooling Load TD For Calculating Cooling Load from Sunlit Walls

North Latitude Wall Facing	1	2	3	4	5	6	7	8	9	10	11	12	13	14	15	16	17	18	19	20	21	22	23	24	Hr of Maximum CLTD	Minimum CLTD	Maximum CLTD	Difference CLTD
													Solar Time, hr															
Group A Walls																												
N	14	14	14	13	13	13	12	12	11	11	10	10	10	10	10	11	11	12	12	13	13	14	14	14	2	10	14	4
NE	19	19	19	18	17	17	16	15	15	15	15	15	16	16	17	18	18	19	19	20	20	20	20	20	22	15	20	5
E	24	24	23	23	22	21	20	20	19	18	18	18	18	19	19	20	21	22	23	24	24	25	25	25	22	18	25	7
SE	24	23	23	22	22	21	20	20	19	19	18	18	18	18	19	19	20	21	22	23	24	24	24	24	22	18	24	6
S	20	20	20	19	19	18	18	17	16	15	14	14	14	14	15	16	17	18	18	19	20	20	20	20	23	14	20	6
SW	25	25	25	24	24	23	22	22	21	20	19	18	17	17	17	17	18	18	19	20	22	23	24	25	24	17	25	8
W	27	27	26	26	25	24	24	23	22	21	20	19	18	18	18	18	18	19	19	20	22	23	25	26	1	18	27	9
NW	21	21	21	20	20	19	19	18	17	16	15	15	14	15	15	15	16	17	18	19	20	20	21	21	1	14	21	7
Group B Walls																												
N	15	14	14	13	13	12	11	11	10	9	9	9	9	10	11	12	13	14	14	15	15	15	15	15	24	8	15	7
NE	19	18	17	16	15	14	13	13	12	12	13	14	16	17	18	19	20	21	21	21	21	20	20	20	21	12	21	9
E	23	22	21	20	20	18	17	16	15	15	15	17	19	22	24	26	26	27	27	26	26	25	24	24	20	15	27	12
SE	23	22	21	20	20	18	17	16	15	14	15	16	18	20	21	23	24	25	26	26	26	26	25	24	21	14	26	12
S	21	20	19	18	17	15	14	13	12	11	11	11	12	13	14	15	17	19	20	21	22	22	21	21	23	11	22	11
SW	27	26	25	24	22	21	19	18	16	15	14	14	14	14	15	17	20	22	25	27	28	28	28	28	24	13	28	15
W	29	28	27	26	24	23	21	19	18	17	16	15	15	15	15	17	19	22	25	27	29	30	30	30	24	14	30	16
NW	23	22	21	20	19	18	17	15	14	13	12	12	12	12	15	15	17	19	21	22	23	23	23	23	24	11	23	9
Group C Walls																												
N	15	14	13	13	12	11	11	10	9	8	8	7	7	8	9	10	11	12	13	14	15	15	16	16	22	7	17	10
NE	19	17	16	15	14	13	13	12	11	10	11	13	15	17	19	20	21	22	23	23	22	21	20	20	20	10	23	13
E	22	21	19	18	17	15	14	12	11	12	14	16	19	22	24	26	28	29	29	29	28	27	26	24	18	12	30	18
SE	22	21	19	18	17	15	14	12	11	12	12	13	15	18	22	24	26	28	29	29	28	27	26	24	19	12	29	17
S	29	27	25	22	20	18	15	13	11	11	11	13	15	18	22	26	29	32	33	33	33	32	31	29	31	11	33	22
SW	31	29	27	25	20	18	15	12	11	10	10	11	12	13	14	16	18	20	24	29	32	33	35	33	22	12	35	23
W	25	23	21	20	18	16	14	13	11	10	10	11	12	13	14	15	16	18	20	22	24	25	26	26	22	10	27	17

(*Continued*)

Table 6-10 (*Continued*)
Cooling Load TD For Calculating Cooling Load from Sunlit Walls

North Latitude Wall Facing	Solar Time, hr 1	2	3	4	5	6	7	8	9	10	11	12	13	14	15	16	17	18	19	20	21	22	23	24	Hr of Maximum CLTD	Minimum CLTD	Maximum CLTD	Difference CLTD
Group D Walls																												
N	15	13	12	10	9	7	6	6	6	6	6	7	8	10	12	13	15	17	18	19	19	19	18	16	21	6	19	13
NE	17	15	13	11	10	8	7	8	10	14	17	20	22	23	23	24	25	25	25	24	23	22	20	18	19	7	25	18
E	19	17	15	13	11	9	8	9	12	17	22	27	30	32	33	33	32	32	31	30	28	26	24	22	16	8	33	25
SE	20	17	15	13	11	10	8	8	10	13	17	22	26	29	31	32	32	32	31	30	28	26	24	22	17	8	32	24
S	19	17	15	13	11	9	8	7	6	7	9	12	16	20	24	27	29	29	29	28	27	26	24	22	19	6	29	23
SW	28	25	22	19	16	14	12	10	9	8	8	8	10	12	16	21	27	32	36	38	38	37	34	31	21	8	38	30
W	31	27	24	21	18	15	13	11	10	9	9	9	10	11	14	18	24	30	36	40	41	40	38	34	21	9	41	32
NW	25	22	19	17	14	12	10	9	8	7	7	8	9	10	12	14	18	22	27	31	32	32	30	27	22	7	32	25
Group E Walls																												
N	12	10	8	7	5	4	3	4	5	6	7	9	11	13	15	17	19	20	21	23	20	18	16	14	20	3	22	19
NE	13	11	9	7	6	4	5	9	15	20	24	25	25	26	26	26	26	25	25	24	22	19	17	15	16	4	26	22
E	14	12	10	8	6	5	6	11	18	26	33	36	38	37	36	34	33	32	30	28	25	22	20	17	13	5	38	33
SE	15	12	10	8	7	5	5	8	12	19	25	31	35	37	37	36	34	33	31	29	26	23	20	17	15	5	37	32
S	15	12	10	8	7	5	4	3	4	5	9	13	19	24	29	32	34	33	31	29	26	23	20	17	17	3	34	31
SW	22	18	15	12	10	8	6	5	5	6	9	12	18	24	32	38	43	43	45	44	40	35	30	26	19	5	45	40
W	25	21	17	14	11	9	7	6	6	6	7	9	13	20	26	34	43	49	49	49	45	40	34	29	20	6	49	43
NW	20	17	14	11	9	7	6	5	5	5	6	8	10	13	16	20	26	32	37	38	36	32	28	24	20	5	38	33

Table 6-11
Cooling Load TD For Calculating Cooling Load From Flat Roofs

Without Suspended Ceiling

Roof No	Description of Construction	Weight lb/ft²	U-value Btu/(h ft² °F)	1	2	3	4	5	6	7	8	9	10	11	12	13	14	15	16	17	18	19	20	21	22	23	24	Hour of Maximum Cltd	Minimum Cltd	Maximum Cltd	Difference Cltd
1	Steel Sheet with 1-in. (or 2-in.) insulation	7	0.213	1	-2	-3	-3	-5	-3	-3	6	19	34	49	61	71	78	79	77	70	59	45	30	18	12	8	5	14	-5	79	84
2	1-in. wood with 1-in. insulation	(8)	(0.124)																												
3	4-in. l.w concrete	8	0.170	6	3	0	1	3	3	2	4	14	27	39	52	62	70	74	74	70	62	51	38	28	20	14	9	16	3	74	77
4	2-in. h.w. concrete with 1-in. (or 2-in.) insulation	18	0.213	9	5	2	0	-2	-3	-3	1	9	20	32	44	55	64	70	73	71	66	57	45	34	25	18	13	16	-3	73	76
5	1-in. wood with 2-in. insulation		(0.122)																												
7	2.5-in. wood with 1-in. insulation	24	0.158	22	17	13	9	6	3	1	1	3	7	15	23	33	43	51	58	62	64	62	57	50	42	35	28	18	1	64	63
8	8-in. l.w concrete	13	0.130	29	24	20	16	13	10	7	6	6	9	13	20	27	34	42	48	53	55	56	54	49	44	39	34	19	6	56	50
9	4-in. h.w. concrete with 1-in. (Or 2-in.) insulation	31 (52)	0.126 (0.120)	35	30	26	22	18	14	11	9	7	7	9	13	19	25	33	39	46	50	53	54	53	49	45	40	20	7	54	47
10	2.5-in. wood with 2-in. insulation	13	0.093	30	26	23	19	16	13	10	9	8	9	13	17	23	29	36	41	46	49	51	50	47	43	39	35	19	8	51	43
11	Roof terrace system	75	0.106	34	31	28	25	22	19	16	14	13	13	15	18	22	26	31	36	40	44	45	43	40	37			20	13	46	33
12	6-in. h.w. concrete with 1-in. (or 2-in.) insulation	75	(0.117)																												
13	4-in. wood with 1-in. (or 2-in.) insulation	(18)	(0.078)																												

(Solar Time, hr. The "Hour of Maximum Cltd", "Minimum Cltd", "Maximum Cltd", and "Difference Cltd" columns appear at right.)

COMPUTERIZED ENERGY CALCULATIONS

There are various routes to speeding up energy and cost calculations in retrofit or new HVAC projects with computers.

1. A potpourri of the formulas needed for HVAC work, yearly energy consumption and costs, ductwork and piping sizing electrical calculations, testing and balancing, air flow, hydronics, fans, pumps etc. is available from Wendes Engineering for $245. This program runs on IBM PC/XT/AT and IBM compatible equipment, TRS 1000/1200 and 2000 computers etc.

2. Spread sheet programs are an excellent approach to evaluating many different systems. R factors and temperature conditions the special engineering programs are easy to apply to spread sheet program and are available from Wendes Engineering.

3. Individual HVAC modules for heating/cooling loads, duct design, piping design, equipment selection etc. can also be purchased.

4. A quick and easy way to perform financial calculations is to buy a dedicated financial calculator which does calculates payback, ROI, PV, FV, interest rates, payments, number of periods, depreciation, amortization, loans, discounted IRR etc. They run between $50 and $120.

5. There are a number of computer energy analysis programs that can be run through telephone modems or by the computer company from the information you give them. They compare the energy savings for different types of systems, energy savings methods etc.

   Accuracy on remote energy analysis systems are with 10 to 15%. Sometimes they can be 20 to 50% off.

6. The computer programs imply various approaches of energy analysis:
   a. Per type of system evaluation.
   b. The bin approach of 5 degree increments.
   c. A complete per hour energy evaluation.

BIN METHOD

The bin method of estimating yearly energy requirements is based on number of hours in five degree increments. One could find, for instance, that an average winter's day had one hour in the zero bin, 4 hours in the 5 bin, 8 hours in the 10 bin, and 11 hours in the 15 bin. Seasonal requirements are determined by adding the total number of hours in each of the 5 bins. (See Table 6-13, page 109, for Bin Method of Calculating Cooling Energy Requirements.)

Table 6-12
Bin Method For Calculating Seasonal Heating Requirements

1 Mean D.B. Temp	2 No. Hours	3 Heat Loss Rate $\times 10^6$	4 Internal Gain $\times 10^6$	5 Setback Bin Factor	6 Heat Loss $\times 10^6$
62	695	.465	.66	.313	—
57	633	.756	.66	.577	—
52	592	1.047	.66	.694	39.44
47	566	1.337	.66	.761	202.32
42	595	1.628	.60	.804	386.10
37	808	1.919	.66	.833	758.33
32	884	2.209	.66	.855	1086.17
27	618	2.500	.66	.872	939.36
22	377	2.791	.66	.885	682.30
17	248	3.082	.66	.896	521.17
12	131	3.372	.66	.905	313.31
7	61	3.663	.66	.913	163 74
2	17	3.954	.66	.919	50.55
-3	4	4.244	.66	.925	13.00
-8	1	4.535	.66	.929	3.55
				TOTAL	5,159.48

Assuming 80% efficiency, input energy = $6,450 \times 10^8$ Btu

HOUR-BY-HOUR CALCULATIONS

Still another, and more exact method, uses basically the same equations already employed to calculate the heat loss in our building. However, these equations are modified by weighting factors that reflect the thermal storage capacities of various construction materials and the characteristics of the energy distribution system. A computer program is used

Table 6-13
Bin Method of Calculating
Cooling Energy Requirements

1 D.B. Temp	2 No. Hours	3 W.B. Temp	4 Sensible Gain $\times 10^6$	5 Solar Gain $\times 10^6$	6 Internal Gain $\times 10^6$	7 W_o	8 Latent Gain $\times 10^6$	9 Load $\times 10^6$
92	33	75	31.8			.0148	11.36	
87	115	70	78.3			*		
82	261	68	103.7			*		
77	456	66	51.7			*		
72	669	62	-113.9			*		
67	791	59	-359.0			*		
62	690	56	-508.9			*		
TOTAL	3015		-716.3	934.7	1992.9		11.36	2,222.7

Table 6-12
Hourly Weather Occurrences

Location	72	67	62	57	52	47	42	37	32	27	22	17	12	7	2	-3	-8	-13	-18
Albany, NY	588	733	740	708	652	625	647	769	793	574	404	278	184	110	63	32	10	5	4
Albuquerque, NM	767	831	719	651	687	734	741	689	552	346	154	66	21	4	1	1			
Atlanta, GA	1185	926	823	784	735	676	598	468	271	112	44	19	8	2					
Bakersfield, CA	831	898	966	977	908	746	541	247	77	7									
Birmingham, AL	1138	908	805	742	668	614	528	433	292	143	69	17	6	3					
Bismark, ND	454	566	614	606	563	520	518	604	653	550	474	371	338	292	278	208	131	77	80
Boise, ID	492	575	643	702	786	798	878	829	522	307	148	53	26	14	6	2	1		
Boston, MA	676	819	804	781	766	757	828	848	674	429	256	151	74	35	4	9	2		
Buffalo, NY	646	772	760	700	666	624	647	756	849	602	426	267	170	815		24			
Burlington, VT	573	670	703	694	655	603	637	716	752	561	491	336	272	216	135	81	39	17	8
Casper, WY	423	532	592	642	606	670	782	831	806	683	495	325	200	116	73	45	30	15	5
Charleston, SC	1267	1090	889	787	651	576	434	321	192	79	27	5							
Charleston, WV	912	949	767	689	661	667	607	633	630	356	252	135	73	22	7	1			
Charlotte, NC	1115	908	839	752	730	684	634	515	360	166	64	23	5	2					
Chattanooga, TN	1021	895	775	722	713	679	642	553	414	228	113	45	4	4	2				
Chicago, IL	762	769	653	592	569	543	591	800	822	551	335	196	117	85	59	25	12		
Cincinnati, OH	879	843	726	639	611	599	627	698	711	460	249	131	68	44	18	8	2	3	
Cleveland, OH	763	831	723	641	638	607	620	754	806	578	355	201	111	47	22	11	2		1
Columbus, OH	774	820	720	648	622	603	658	730	772	502	280	169	94	40	20	10	4		
Corpus Christi, TX	1175	1041	748	551	444	302	180	83	27	9	3								

(Continued)

Table 6-12 (Continued)
Hourly Weather Occurrences

Location	72	67	62	57	52	47	42	37	32	27	22	17	12	7	2	-3	-8	-13	-18
Dallas, TX	831	795	693	656	629	576	504	371	231	91	34	17	4	1	36	22	6	1	1
Denver, CO	549	684	783	731	678	704	692	717	721	553	359	216	119	78	104	59	23	8	1
Des Moines, IA	707	751	681	600	585	512	510	627	747	557	405	281	211	152	17	4	1		
Detroit, MI	721	783	695	633	592	566	595	808	884	618	377	248	131	61					
El Paso, TX	933	839	749	760	687	611	494	369	233	104	34	10	2						
Fort Wayne, IN	728	777	699	608	569	552	601	725	905	596	381	205	124	69	40	19	6	1	
Fresno, CA	709	803	921	1006	1036	952	673	426	168	34									
Grand Rapids, MI	634	739	712	647	571	565	554	742	938	690	469	293	172	78	31	10	1	1	
Great Falls, MT	407	520	636	754	822	830	832	813	698	533	355	218	167	136	118	101	68	51	62
Harrisburg, PA	807	824	737	692	635	659	722	888	749	427	222	125	52	18	4	1			
Hartford, CT	617	755	751	752	749	575	683	807	825	552	370	233	153	77	33	11	3	2	
Houston, TX	1172	980	772	681	570	452	291	141	64	18	4	2							
Indianapolis, IN	821	815	722	585	586	579	605	712	791	551	293	152	97	60	35	13	3	2	
Jackson, MS	1168	922	790	677	618	605	484	367	224	103	41	6	2	2	1				
Jacksonville, FL	1334	975	879	692	530	355	288	154	83	24	2								
Kansas City, MO	761	723	601	572	553	562	628	625	591	407	265	175	99	51	21	4			
Knoxville, TN	1056	889	746	675	672	689	648	590	456	217	101	41	21	7	2				
Las Vegas, NV	651	644	699	786	769	716	591	396	194	44	7	1							
Little Rock, AR	940	803	725	672	638	669	605	509	363	172	50	25	5	1					
Los Angeles, CA	881	1654	2193	1904	1054	428	107	10											
Louisville, KY	869	758	693	654	619	634	649	703	631	332	169	97	45	25	8	3	1		
Lubbock, TX	833	829	688	700	642	618	620	546	490	346	180	86	33	7	5	1			
Memphis, TN	977	798	715	690	618	633	614	532	374	196	74	25	10	4					
Miami, FL	1705	810	452	277	147	71	26	4											
Milwaukee, WI	597	753	749	634	585	591	611	774	913	659	421	285	176	116	83	47	18	4	1

City																		
Minneapolis, MN	621	690	695	602	588	482	500	560	632	609	514	383	311	246	186	119	62	31	16
Mobile, AL	1411	1038	882	698	609	506	377	214	109	49	7	3			3	1	1	1	
Nashville, TN	933	838	738	697	637	619	627	565	463	263	132	67	28	9					
New Orleans, LA	1189	987	850	692	621	449	282	128	47	9	2				1				
New York, NY	926	877	754	745	722	796	838	858	603	330	188	2	26	10		1			
Oklahoma, OK	881	769	717	643	645	611	641	570	468	287	173	77	36	12	3	1			
Omaha, NB	726	721	606	558	539	543	543	655	663	511	390	387	189	135	93	40	15	1	1
Philadelphia, PA	863	809	735	710	663	701	758	818	654	335	189	100	32	9					
Phoenix, AZ	762	776	767	769	659	540	391	182	57	8									
Pittsburg, PA	722	910	799	678	637	587	631	688	569	774	360	233	159	60	30	7	1		
Portland, ME	407	627	780	808	760	748	722	839	820	599	408	293	190	109	60	29	15	5	1
Portland, OR	373	581	1001	1316	1274	1271	1238	772	343	123	40	10	4	1					
Raleigh, NC	1087	937	848	762	707	672	638	527	410	236	103	38	11	1	15	4	1		
Reno, NV	418	477	572	690	845	909	890	829	733	530	387	277	101	37	1				
Richmond, VA	953	850	784	745	690	673	699	632	478	285	138	67	19	2					
Sacramento, CA	630	773	1071	1329	1298	1049	701	355	93	8									
Salt Lake City, UT	569	615	614	635	682	685	755	831	798	564	328	158	80	41	16	2			
San Antonio, TX	1086	943	789	669	569	445	387	190	94	31	11	4	1	1					
San Francisco, CA	285	665	1264	2341	2341	1153	449	99	10										
Seattle, WA	258	448	750	1272	1462	1445	1408	914	427	104	39	20	3						
Shreveport, LA	1063	886	772	679	619	609	516	361	200	72	23	6	2						
Sioux Falls, SD	566	684	669	605	522	498	501	625	712	585	520	448	293	208	152	102	59	43	18
St. Louis, MO	823	728	646	575	585	578	620	671	650	411	219	134	77	40	15	7	1	2	2
Syracuse, NY	627	735	723	717	656	641	651	720	830	547	392	282	190	102	55	23	5		
Tampa, FL	1387	1187	877	570	345	216	137	48	10	1									
Waco, TX	909	830	701	622	651	558	501	354	316	84	24	3	1						
Washington, DC	960	766	740	673	690	684	790	744	542	254	138	54	17	2	14	3			
Wichita, KS	758	709	641	603	589	592	611	584	607	426	273	161	85	45					

Chapter 7

Financial Evaluations

The purpose of the following financial evaluations is to determine payback periods, return on investments and rates of return. This involves first costs, operation and maintenance costs, depreciation, taxes, energy savings, present values, future values, interest rates, time periods and salvage values.

Example Retrofit Investment Evaluation

Retrofit costs (first costs)	$20,000
O & M per year	1,000
Depreciation per year (straight line 10 years)	2,000
Savings on taxes	2,000
Savings on energy per year	10,000

1. **Simple Payback**

$$\text{Payback} = \frac{\text{First Costs}}{(\text{Annual Energy Savings}) - (\text{Yrly O\&M Costs})}$$

$$\text{Payback} = \frac{\$20,000}{\$10,000 - \$1,000} = 2.22 \text{ years}$$

2. **R.O.I. Return On Investment**

$$\text{R.O.I.} = \frac{\dfrac{\text{Annual Energy}}{\text{Savings}} - \dfrac{\text{Depreciation}}{\text{Cost Per Yr}} - \dfrac{\text{O \& M Cost}}{\text{Per Year}} - \dfrac{\text{Savings}}{\text{On Taxes}}}{\text{First Costs}}$$

$$R.O.I. = \frac{\$10,000 - \$2,000 - \$1,000 - \$1,000}{\$20,000} = 30\%$$

3.   Present Value

With constant savings per period use **discount** and sum of benefits during expected life of equipment.

Use present worth factor (PWF) from compound interest tables, using known discount rate and life of investment.

Multiplying the annual savings (S) by PWF will equal the PV of future savings.

For example $1000 invested at 8% compound interest for 5 years will be worth $1469 in 5 years.

$680 invested today will be worth $1000 in 5 years. This can be stated that the present worth of $1000 five years from now is only $680.

Chart 7-1
Payback Analysis Chart

Cash
Investment
divided by
Savings

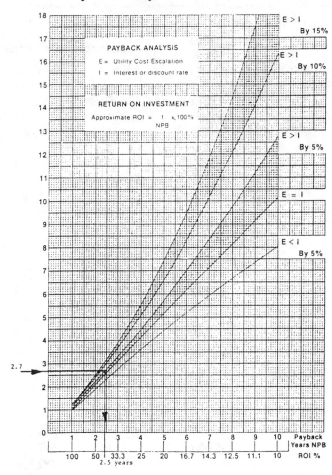

PAYBACK ANALYSIS

E = Utility Cost Escalation
I = Interest or discount rate

RETURN ON INVESTMENT

Approximate ROI = $\frac{1}{NPB} \times 100\%$

E > I
By 15%

E > I
By 10%

E > I
By 5%

E = I

E < I
By 5%

**Payback Period and
Return on Investment**
Capital Investment C = $29,263.00
Savings Utility Cost (1st years)
    S = $10,770.00
Utility Cost Escalation E = 15%
Interest I = 10%
E > I by 5%
E < I by _____ %

$$\frac{\text{Capital Investment}}{\text{Savings 1st Yr.}} = \frac{C}{S} = \frac{29,263.00}{10,770.00} = 2.7$$

From Payback Curves:
    Payback Period N$_{PB}$ = 2.5 years
Approximate Return on Investment
    ROI = 1 ÷ N$_{PS}$ = 1 = 40%

Factors Affecting
Owning Costs
    Initial Costs
    Interest
    Time Period
        Useful life
        Depreciation period
    Present Worth
    Property Tax
    Insurance
    Salvage Value

Factors Affecting
Operating Costs
    Energy Costs
    Cost Escalation
    Maintenance
    Labor Operation
    Water Costs

Table 7-1
Future Value Of Energy Savings

Net Dollar Savings Per Month	Net Dollar Energy Savings Per Yr	Compounded $ Value at 10% Interest		
		Period		
		5 Year	10 Year	15 Year
$8.33	$100	$645	$1,706	$3,451
41.67	500	3,227	8,534	17,265
83.33	1,000	6,452	17,065	34,525
166.67	2,000	12,904	34,133	69,054
250.00	3,000	19,357	51,199	103,579
333.33	4,000	25,809	68,266	138,108
416.67	5,000	32,262	85,860	172,633
$500	$6,000	$38,714	$102,399	$207,158
583	7,000	45,166	119,464	241,683
667	8,000	51,619	136,532	276,212
750	9,000	58,072	153,597	310,737
833	10,000	64,524	170,664	345,262
$1,000	$12,000	$77,429	$204,798	$414,316
1,167	14,000	90,334	238,928	483,370
1,333	16,000	103,238	263,064	552,420
1,500	18,000	116,143	207,194	621,474
1,667	20, 000	129,048	341,328	690,528
$2,083	$25,000	$161,309	$429,300	$863,157
2,500	30,000	193,572	511,992	1,035,796
2,917	35,000	225,834	597,326	1,208,426
3,333	40,000	358,096	682,656	1,381,056
3,750	45,000	290,357	767,989	1,553,686
$4,167	$50,000	$325,000	$853,322	$1,726,324
5,000	60,000	387,143	1,023,990	2,071,592
5,833	70,000	451,667	1,194,650	2,416,842
6,667	80,000	516,191	1,365,320	2,762,112
7,500	90,000	580,715	1,535,980	3,107,372
8,333	100,000	645,238	1,706,640	3,452,620

Net Energy Savings Per Year Equals Total Energy Savings Less:
- O & M Costs
- Depreciation
- Taxes

Table 7-2
Summary of Lighting Costs for Lighting Reduction
In A 48,000 Sq Ft Area

Design Options	Lighting kW	Foot candles[1]	Lighting kWh (1000)	Power cost (dollars)[2]	Cost difference from original	Cost difference from A2	implementation cost Lamp (dollars)	Labor (dollars)	Total installation cost (dollars)	Payback years
Original design	269.7	200	2,363	135,375	—	—	Present practice			
A-2 present practice	229.2	171	2,008	115,046	20,329	—	Present practice			
A-3	141.9	122	1,243	71,226	64,149	43,820	23.159	10,050[4]	33,209	0.76
A-4[3]	98.2	92	860	49,291	86,084	65,755	23,728	10,050[4]	33,778	0.51
A-5	114.7	85	1,005	57,574	77,801	56,472	—	10,050[5]	10,058	0.17

RATE OF RETURN, LIFE CYCLE DISCOUNT

The Rate of Return Cost Analysis is based on life cycle cost analysis. Life cycle cost analysis is a useful approach because it takes into account all costs over the life of a project rather than first costs only. These costs include costs of acquisition, maintenance, operation, and, where applicable, disposal.

The evaluation of the economic viability of a project should be based on total cost or cost savings, discounted to reflect the time values of monetary amounts that are expected to accrue over the useful life of the project.

To compare money through time, it is necessary to convert monetary amounts to an equivalent base. This is done by applying appropriate discounting formulas to the monetary amounts for conversion to either a present value basis or an annual cost basis.

Theoretically, the-rate of return (ROR) method of life cycle costing is based on the observation that the return on any project is used for just two purposes:

• To repay all costs (capital investment as well as expense).

• To pay a return on unrecovered capital. Thus, the method's objective is to determine the rate or -return on funds that are actually "tied up" in a project.

The rate of return that will be realized is calculated. If only one compound interest factor is involved in the cash flow equation, the problem can be solved directly; when two or more factors are involved, the method of solution becomes one of trial-and-error.

The rate of return may also be defined as the interest rate on unrecovered investment balances where the payment schedule causes the unrecovered investment to equal zero at the end of the life of the investment. The net end-of-year unrecovered investment balance can thus be determined.

The net investment balances are negative each year, indicating that the project requires more money than it generates. In other words, the project is continually "in debt" until the end of the last year when a final payment is made and the net investment balance becomes zero. The firm has recovered exactly the amount of invested funds along with interest at X percent on the unrecovered investment balances. Thus, it has been shown that the return on the energy conservation measure provides for the return of and on unrecovered capital.

Section 4

Retrofit Estimate Costs

Chapter 8

Budget Estimating HVAC Energy Retrofit Costs and Estimating Procedures

Chapter 9
Sample Energy Retrofit Estimate

Budget Estimating HVAC Retrofit Costs and Estimating Procedures

This chapter covers budget estimating costs for HVAC retrofit work, comparing retrofit costs to savings, sample chiller retrofit costs and the procedure for preparing HVAC retrofit estimate.

COSTS TO MOVE AND REPLACE

1.	New Pumps	$4 to $15/gpm
2.	New Fan Drives & Motors	$150/HP+
3.	New Chillers	$300 to $450/ton
4.	New Boilers	$8 to $14/Mbh
5.	Heat Recovery Installations	$1 to $4.50/Cfm
6.	Rebalance	$14 to $20 outlet $80 to $180 per AHU coil $30 per terminal unit
7.	Convert CAV to VAV	$600 to $1800/box $.50 to $1.50/sq ft
8.	Audits	$3 to $8/sq ft
9.	Overall Energy Retrofit Programs	$.35 to $6.00/sq ft

Table 8-1
Budget Estimating Energy Retrofit Costs
Compared to BTU and Costs Savings

Retrofit Investment Costs Per Sq Ft Building	Rough Yearly Btus Saved Per Sq Ft	Yearly Energy Costs Saved Dollars		ROI Percent In Decimals
		Per Sq Ft	Per 1,000 Sq Ft	
Office Buildings				
$0.25	40,000	$0.86	$860.00	3.37
0.50	46,000	0.99	990.00	1.91
0.75	50,000	1.08	1,080.00	1.37
1.00	53,000	1.14	1,140.00	1.07
1.25	56,000	1.20	1,200.00	0.89
1.50	58,000	1.25	1,250.00	0.77
1.75	59,000	1.27	1,270.00	0.66
2.00	60,000	1.29	1,290.00	0.58
2.25	60,500	1.30	1,300.00	0.51
2.50	61,000	1.31	1,310.00	0.46
2.75	61,500	1.32	1,320.00	0.41
3.00	62,000	1.33	1,330.00	0.38
Hospitals				
$0.25	45,000	$0.97	$970.00	3.81
0.50	56,000	1.20	1,200.00	2.33
0.75	62,000	1.33	1,330.00	1.71
1.00	65,000	1.40	1,400.00	1.33
1.25	67,000	1.44	1,440.00	1.09
1.50	70,000	1.51	1,510.00	0.94
1.75	72,000	1.55	1,550.00	0.82
2.00	74,000	1.59	1,590.00	0.73
2.25	75,000	1.61	1,610.00	0.65
2.50	76,000	1.63	1,630.00	0 59
2.75	77,000	1.66	1,660.00	0 54
3.00	78,000	1.68	1,680.00	0.49

(Continued)

Table 8-1 (*Continued*)

Retrofit Investment Costs Per Sq Ft Building	Rough Yearly Btus Saved Per Sq Ft	Yearly Energy Costs Saved Dollars		ROI Percent In Decimals
		Per Sq Ft	Per 1,000 Sq Ft	
Universities				
$0.25	30,000	$0.65	$650.00	2.53
0.50	52,000	1.12	1,120.00	2.17
0.75	61,000	1.31	1,310.00	1.68
1.00	67,000	1.44	1,440.00	1.37
1.25	72,000	1.55	1,550.00	1.17
1.50	77,000	1.66	1,660.00	1.04
1.75	80,000	1.72	1,720.00	0.92
2.00	84,000	1.81	1,810.00	0.84
2.25	87,000	1.87	1,870.00	0.76
2.50	90,000	1.94	1,940.00	0.71
2.75	92,000	1.98	1,980.00	0.65
3.00	93,000	2.00	2,000.00	0.60

1. Total savings and ROI based on electrical and fuel consumption.
2. Electrical savings based on 66 percent of total consumption.
3. Gas and oil based on 34 percent of total consumption.

Table 8-2
Budget Estimating Chiller Retrofits
300 Ton Chiller

	Material Costs	Labor Hours	Totals w/30% MU
Chiller @ $300/ton	$90,000	130	$96,500
Piping	$1,500	100	$6,500
Electrical	$15,00	100	$6,500
Insulation	$1,500	100	$6,500
Controls	$1,500	100	$6,500
General Construction	$1,500	100	$6,500
Total	$97,500	630 hr	$129,000

Hence $129,000 divide 300 tons = $430 per ton.

ESTIMATING RETROFIT COSTS

Estimating is calculating labor, material and overhead costs on a project and coming up with a reasonably accurate total price which properly reflects the final actual costs of material and labor of the particular project.

Pricing is frequently based on imperfect or incomplete data, vague and inadequate plans and specs, and unpredictable and uncontrollable conditions creating an element of risk and potential error.

Hence, it is sometimes a relatively difficult, complicated and a judgement type affair and consequently, must be handled with great care using valid price charts, quotations etc.

The procedure involves that the engineering design be completed first, either in complete detail, conceptually or partially designed in scope fashion.

A listing should be made of all types of items required.

Then quantities, types, sizes, accessories, building conditions etc. must all be established and extended for labor and material. Sub-contractor and supplier quotations must be obtained and then all items summarized.

Figure 8-1

Procedure for Preparing A Retrofit Estimate

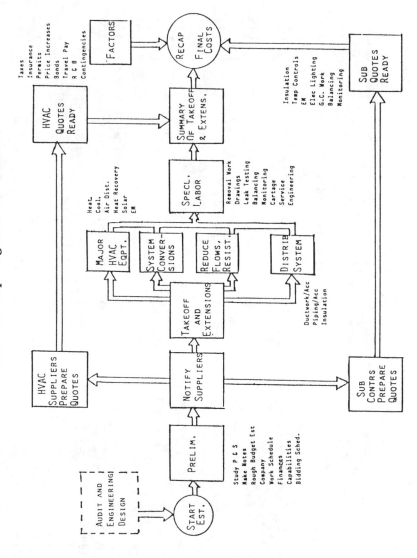

Table 8-3
Field Labor Correction Factors For HVAC Equipment, Ductwork & Piping
Use as Multipliers Against Labor Hours

Floors

Bsmt, ground, 1st	1.00
2nd & 3rd	1.04
4th, 5th, 6th	1.08
7th, 8th, 9th	1.12
10th, 11th, 12th	1.16
13th, 14th, 15th	1.21
16th, 17th, 18th	1.26
19th, 20th, 21st	1.31

Duct Heights

10 ft	1.00
15 ft	1.10
20 ft	1.20
25 ft	1.30
30 ft	1.40
35 ft	1.50

Existing Building

Typical existing office bldg, hospital, school, etc.	1.35
Existing factory, warehouse, gym, hall, garage, no ceilings	1.15
100% gutted area	1.10
Work around, over, machinery, furniture	1.10
Protect machinery, floor furnit.	1.05
Quiet job	1.05
Occupied areas	1.05
Remove items and reinstall same	1.50
Remove and replace equipment	2.00

Cost Control

Piping or ductwork installed	1.10
Electrical conduit installed	1.05

(Continued)

40 ft	1.60

Correction for Size of Job

0 - 24 hrs	1.12
25 - 48 hrs	1.08
48 - 96 hrs	1.04
96 hrs up	none

Special Areas

Open areas, no partitions	.85
Congested ceiling space	1.15
Equipment room	1.20
Kitchen	1.10
Auditorium; pool w/sloped flr	1.25
Attic space	1.50
Crawl space	1.20
Cramped shaft	1.30
One or two continuous risers	.80

Partitions & door frames erected	1.10
Ceiling grid installed	1.05
Overmanning job, crashing	1.20
Company overloaded w/work	1.20
Not being on top of job	1.20
On top of job continually	.90
Go back to put something in	1.15
Out of phase work	1.15
Move stock pile about	1.05
Delays of deliveries, shop drwgs, appvls, purch., fab., tools	1.15
Poor shop drawings	1.15
Poor foreman	1.20
Delay in facing prob. decision	1.15
Cluttered floors	1.15
Trades working on top of ea. other	1.10
No service roads, muddy	1.10
Congested traffic, poor unloading	1.10

(Continued)

Table 8-3 (Continued)

Distance From Unloading Point

100 ft	1.00
200 ft	1.03
300 ft	1.05

Overtime

Overtime	40 to 50 hrs	1.10
	50 to 60 hrs	1.15
Night work	4 to 12 midnight	1.20
	12 to 8 am	1.20

Correction For Temperature

Under 20%	1.15
Over 90%	1.15

Chapter 9

Sample HVAC Retrofit Estimate

The following sample retrofit estimate is based on the sample audit of the Suburban office building shown in chapter 4. There are 14 areas of costs involved which have to be priced up as shown in the proposed renovation list. Detailed estimate sheets are shown on the audit, outside air reduction, fan Cfm reduction, new pumps, the VAV conversion, computer room and at the end there is a final recap sheet.

SAMPLE ENERGY RETROFIT ESTIMATE

Proposed Energy Savings Renovations

1. Perform a thorough detailed audit with complete test readings, engineering calculations and pricing, after which decisions are made on what retrofit work will be estimated and performed.

2. The reheat system will be converted to variable air volume (VAV). Some reheat coils will be eliminated. An energy management system will be installed to automatically control fan volume etc.

3. The induction system fan will be turned off in winter and run at a reduced rate only during occupancy during summer.

4. Lighting levels will be reduced from 4 watts per sq. ft to

5. Winter space thermostat settings to be reset from 74°F to 70°F and summer from 74°F to 78°F. All stats to be checked and recalibrated if needed.

6. Fan Cfm's will be reduced by reducing RPM's.

7. Pump GPM's will be reduced by replacing pump.

8. Minimum outside ventilation air will be reduced from 30,000 Cfm to 13,000 Cfm.

9. All filters, coils, strainers, condenser tubing and boiler tubing to be checked and cleaned as required.

10. Change computer room unit from a CHW to DX system and reduce the tonnage from 26 to 18 tons. Also add a economizer section on to the unit and include staging in the condenser.

11. The boiler will be cleaned, checked and tuned up.

12. Install computer controlled optimizer on chiller.

13. Install automatic combustion controls and temperature reset controls on boiler.

14. An energy management system will be installed to control on and off times of lighting, fans, HVAC units, pumps, chillers and to minimize demand costs.

15. The systems will be rebalanced, air and water, for maximum efficiency of operation and optimum minimum energy consumption.

ESTIMATE SUMMARY & EXTENSION SHEET

Date ___11-18-92___

Page _____

Job ___Suburban Office Bldg___

DETAILED AUDIT HOURS

Qty.	Field Inspection and Tests:*	MATERIAL COST		LABOR			
		UNIT	TOTAL	UNIT	TOTAL	UNIT	TOTAL
8	Test Fans,			2	16		
3	Test Pumps			2	6		
1	Test Boiler				3		
	Test Chiller				3		
	Test Cooling Tower				6		
5	Check Strainers			3/4	4		
22	Spot Check Reheat Boxes, 20%			1/4	11		
16	Spot Check Induction Units, 20%			1/4	4		
36	Spot Check Thermostats, 20%			1/4	9		
	Spot Check Control Valves				4		
	Spot Check Control Dampers				4		
2	Check Exhaust Systems				2		
	Lighting Survey				4		
	TOTAL				76	HR	
	Engineering:						
	Study Plans				2		
	Evaluation of Energy Use				8		
	Calculate Potential Energy Savings				8		
	Calculate Pay Backs				4		
	Budget Retrofit Costs				6		
	Prepare Recommendations				6		
					36	HR	
	Total Hours				112	HR	
					$40	HR	
	Total				$4800		
	* Flows, Suction and Discharge Pressures,						
	RPM's, Amps, Volts, Temperatures						

ESTIMATE SUMMARY & EXTENSION SHEET

Date 11-18-92

Page _____

Job SUBURBAN OFFICE BLDG

REDUCE MINIMUM OUTSIDE AIR QUANTITIES

		MATERIAL COST		LABOR			
		UNIT	TOTAL	UNIT	TOTAL	UNIT	TOTAL
	REDUCTIONS:		REDUCTIONS				
	S-1 7,500 TO 2500 CFM		5000	CFM			
	S-2 7,500 TO 2500 CFM		5000	CFM			
	S-3 7,000 TO 5000 CFM		2000	CFM			
	S-4 2,000 TO 1000 CFM		1000	CFM			
	S-5 4,000 TO 1000 CFM		3000	CFM			
	S-6 2,000 TO 1000 CFM		1000	CFM			
	OLD 30,000 13000 NEW		17000	CFM			
6	REMOVE DAMPERS, SUPPLY			2	12		
6	NEW DAMPERS, 15 SQ FT EA AT $15/	$225	1350	5	30		
6	NEW LINKAGES	20	120	1	6		
4	CHANGE DUCTS, 200 LB	1.50	300	3	12		
12	CK, ADJUST, LUBRICATE, SEAL	20	120	1	12		
	RA & EXH DAMP. (REUSE)		---		---		
18	TEST, SET			1	18		
	TOTALS		$1890		90		
	30% O & P		570	a	$49		
			$2460		$3600		
	TOTAL MATERIAL & LABOR		$6060				

ESTIMATE SUMMARY & EXTENSION SHEET

Date 11-18-92

Page _____

Job SUBURBAN OFFICE BLDG

REDUCE CFM OF FANS

		MATERIAL COST		LABOR			
		UNIT	TOTAL	UNIT	TOTAL	UNIT	TOTAL
	REDUCE 30%:						
	S-1 38,000 TO 26,000 CFM						
	S-2 38,000 TO 26,000 CFM						
	R-1 32,000 TO 22,000 CFM						
	R-2 32,000 TO 22,000 CFM						
	(S-3 INDUCT 7000 TO 5000 CFM)						
	30% VOLUME DECREASES						
	HP 49%						
2	REMOVE 75 HP MOTORS			10	20		
2	REMOVE 15 HP MOTORS			3	6		
2	NEW 25 HP MOTORS	768	1,536	11	22		
2	NEW 7½ HP MOTORS	320	640	5	10		
2	INSTALL 25 HP DRIVES	130	260	3	6		
2	INSTALL 7½ HP DRIVES	110	220	2	4		
4	STARTUP, TEST			2	8		
	TOTALS		$2,656		76		
	30% O & P		797	a	$40		
			$3,453		$3,040		
	TOTAL MATERIAL & LABOR		$6,493				

ESTIMATE SUMMARY & EXTENSION SHEET

Date 11-18-92

Page _____

Job SUBURBAN OFFICE BLDG

INSTALL NEW PUMPS

	30% REDUCTION IN FLOW REDUCE PUMPS BHP FROM 25 TO 5 BHP 744 GPM TO 459 GPM	MATERIAL COST		LABOR		INSTALLED COSTS WITH O & P	
		UNIT	TOTAL	UNIT	TOTAL	UNIT	TOTAL
2	REMOVE 12" DIA. 30 HP PUMPS			9	18		
2	NEW 8" DIA. 10 HP	2218	4436	16	32		
	MISC. PIPING, VALVES		500		16		
2	STARTUP, TEST OUT			3	6		
	TOTALS		4936		72		
	30% O & P		1481	@	$40		
			$6417		$2880		
	TOTAL MATERIAL & LAB		$9297				

ESTIMATE SUMMARY & EXTENSION SHEET

Date __11-18-92__

Page _____

Job __SUBURBAN OFFICE BLDG__

		VAV CONVERSION

		MATERIAL COST		LABOR			
		UNIT	TOTAL	UNIT	TOTAL	UNIT	TOTAL
76	REMOVE CAV BOX CONTROLLERS			1	76		
76	NEW VAV BOX CONTROLLERS	250	19,000	2	152		
	FLOW MEASURING STATIONS:						
2	S-1, S-2 72 x 24	777	1,554	4.3	8.6		
2	R-1, R-2 60 x 36	585	1,170	3.5	7		
2	STATIC TOTALIZERS	150	300	2	4		
2	VOLUMETRIC CONTROL CENTERS	700	14,000	3	6		
2	INLET VANE DAMPERS, 33" DIA.	145	290	2.6	5.2		
2	INLET VANE DAMPERS, 36" DIA.	163	326	2.8	5.6		
4	CONTROL MOTORS ETC.	400	1,600	2	8		
12	CONTROL TUBING HOOKUP		2,000	4	48		
1	MOTOR SPEED CONTROLLER, 25 HP		4,700		13		
10	DUCTWORK CHANGES	100	1,000	3	30		
	ELECTRICAL WIRING, AVG. 30 FT	10	300				
	FIELD MEASURE				24		
	PROTECT FURNISHINGS				8		
	SETUP, CLEAN UP				8		
	PROBLEMS, COORDINATION				8		
	TOTALS		$46,000		412	HR	
	30% O & P		13,800	a	$40		
			$59,800		16,480		
	TOTAL MATERIAL & LAB		$76,300				
	CLOSE REHEAT COIL VALVES EXCEPT EVERY FORTH ONE						

ESTIMATE SUMMARY & EXTENSION SHEET

Date __11-18-92__

Page _____

Job __SUBURBAN OFFICE BLDG__

CHANGE COMPUTER RM. UNIT TO DX

	REDUCE FROM 26 TONS TO 18 TONS	MATERIAL COST		LABOR			
		UNIT	TOTAL	UNIT	TOTAL	UNIT	TOTAL
1	REMOVE CHILLED WATER COIL				16		
1	NEW DX COIL, 18 TONS		780		16		
	PIPING & VALVES		500		16		
1	NEW CONDENSER UNIT 18 TONS		6000		16		
	ELECTRIC WIRING, 70 FT		500		8		
	CONTROL WIRING, 50 FT		150		4		
	STRUCTURAL STEEL ROOF SUPPORT		120		4		
	CUT & SEAL ROOF OPENING		20		3		
	MEASURE				4		
	STARTUP				4		
	TOTALS		$8070		91		
	30% O & P		2421	@	$40		
			$10,491		$3640		
	TOTAL MATERIAL & LABOR		$14,131				

ESTIMATE SUMMARY & EXTENSION SHEET

Date 11-18-92

Page _____

Job SUBURBAN OFFICE BLDG
90,000 SQ FT

RECAP

	DIRECT MATERIAL COST		FIELD HOURS		TOTAL COSTS WITH O & P	
	UNIT	TOTAL	UNIT	TOTAL	UNIT	TOTAL
DETAILED AUDIT		---		112		$4,800
REDUCE OUTSIDE AIR QUANTITIES		$1,890		90		6,060
REDUCE LIGHTING LOAD				140		5,600
REPLACE THERMOSTATS, 12		1,800		24		2,772
REDUCE CFM OF FANS		2,656		76		6,493
INSTALL SMALLER PUMPS		4,936		72		9,297
VAV CONVERSION		46,000		412		76,300
REVAMP INDUCTION SYSTEM		1,000		16		1,940
BOILER, COMBUSTION CONTROLS		3,000		20		4,700
CHANGE COMPUTER ROOM AHU TO DX		8,070		91		14,131
INSTALL EMS UNIT		33,600		707		71,960
BOILER STACK HEAT RECOVERY		8,020		101		14,466
REBALANCE		---		244		9,760
TOTALS		$110,972		2,105		
30% O & P		33,292	a	$40		
		$144,264		$84,160		$228,424

Section 5

Reduce Heating and Cooling Loads

Chapter 10

Reduce Heating and Cooling Loads

Reducing heating and cooling loads involves changing space temperature settings, lighting reductions, turning equipment and lights off when not occupied, reducing minimum outside air, using free cooling, reducing heating cooling factors of structure and controlling stratification of air in spaces Controlling outside air correctly is more difficult task and there are potential problems. This area of load reduction is covered extensively in this chapter.

Change Space Temperature Settings
- Turn thermostats down to 70°F in winter.
- Turn thermostats up to 75°F in summer.
- Set stats back or set about 10°F evenings and weekends.
- Do manually, with individual set back stats or with E.M.S.

Reduce Amount of Lighting
- Reduce lighting levels. Delamp.

Reduce Minimum Outside Air Quantities

Turn HVAC Equipment off When Not Occupied
- Shut down HVAC equipment during non-occupied hours, nights, weekends, etc. Put into effect on a manual basis or install small, modular E.M.S. units.

- Turn domestic hot water pumps off during non-occupied hours, nights, weekends, etc.

- Turn off all toilet, kitchen and lab exhaust fans during non-occupancy, nights, weekends.

- Shut down make-up air units and HVAC zones when unoccupied.

Turn Lighting off When Not Occupied
- Turn off lighting during non-occupied hours, nights, weekends, etc. Execute on a manual basis or install timers or small modular E.M.S. units.

- Shade windows during summer.

- Draw drapes, curtains, blinds to reduce solar loads during summer.

Turn Heat Producing Equipment off When Not in Use

Reduce Structural Transmission, Solar and Infiltration Gains and Losses
- Provide maximum economical insulation.
- Double or triple glaze windows.
- Seal infiltration leaks.
- Provide window protection against solar radiation summer.

Use Free Energy
- Use outdoor air for free cooling when O.A. temperatures are roughly between 40°F and 60°F during cooling season.

- Use outdoor air for free cooling during winter when outside air temperatures are below 60°F in spaces that require cooling year round.

- Where economizers exist, make sure they operate properly.

Reduce Absortivity of Walls and Roof with Lighter Colors.

Control Temperature Stratification of Air in Spaces with Ceiling Propeller Fans, etc.

KEEP IT MAINTAINED

HVAC Equipment Maintenance
- Keep filters, strainers, traps, coils, condenser tubes, boiler tubes etc. clean to minimize resistance and maximize heat transfer.

- Check and adjust flows, rpms, amp draws of fans, pumps, chillers and adjust for maximum efficiency.

- Make sure everything is sealed.

- Make sure everything is lubricated.

- Check drives for proper tension alignment, wear, correct power transmission, etc.

Controls Maintenance
- Check calibrations on thermostat, aquastats, pressure stats, etc. and correct as required.

- Check operation of control motors for dampers and valves to see that they operate correctly through full cycle, the linkages are fastened correctly and make sure that there is no binding either in the damper or valves as well as with the linkages.

- Make sure economizer cycles operate properly.

OUTSIDE AIR ENERGY SAVINGS

Energy is wasted when the temperature of the minimum outside air, or mixed air is raised or lowered to required temperature levels. The general goal of outside air energy savings is not to pay a penalty for reheating or recooling excessive outside air quantities due to incorrect minimums, damper leakage, poor control, incorrect control settings, etc.

Correctly and consistently controlled minimum outside air volumes, low damper leakage, shutting outside air off when not needed, using outside air for cooling when cooling is required can all be achieved as follows.

Check out Actual Outside Air Conditions
1. Measure actual amount of outside air being taken in under different outside temperature conditions and building load conditions. Check if minimum air volumes are correct and being held, and if sufficient amounts of cool outside air is being taken in when required for cooling.

Adjust damper linkages as required and reset controls or change controls as required.

2. Check damper leakage when closed. Check different percentage damper openings to see if the percentage of actual air volume matches the percentage damper setting, throughout the damper opening and closing cycle. Install sealing strips on dampers if this will work, if not install new low leakage dampers.

Reduce to Minimum Outside Air Needed

3. Reduce to the most economical amount of minimum outside air when not needed for cooling while still meeting code, healthy ventilation, building pressure and direct exhaust makeup air requirements.

    Rule of thumb for outside air requirements is 10 to 15 percent of total Cfm. For example a 10,000 Cfm system for a 10,000 sq. ft office building would need a minimum of 1000 to 1500 Cfm.

    This equates to .10 or .15 Cfm per sq ft of building.

    In terms of building occupants it equates to:

    @ 1 person per 100 sq ft
    = 100 people @ 10 Cfm ea = 1000 Cfm
    = 100 people @ 15 Cfm ea = 1500 Cfm

Use Effective Minimum Outside Air Controls

4. Control minimum amount of outside air accurately and consistently.

    The typical older control system with a constant mixed air temperature controller using a sensor in the mixed air plenum does not control the volume well, because it generally exceeds minimum OA requirements. The percentage of outside air coming in under the minimum mode may vary from 10% to 100% depending on the OA temperature.

    For example, the following percentages of OA are required to maintain a fixed 55°F mixed air temperature with a constant 70°F return temperature with a constant mixed air temperature controller.

Outside Air Temperature, °F	Percent OA Required to Maintain 55°F
–10°F	19%
0	21
10	25
20	30
30	37
40	50
50	75
55	100

The average percentage of outside air intake during the winter season for various cities is:

Chicago, 42 percent
Boston,   47 percent
Denver,   44 percent
Atlanta, 54 percent

5. The solution to this problem of the constant mixed air controlling not controlling the volume of outside air correctly is to move the constant mixed air temperature controller to the leaving side of the heating coil and use it as a low limit temperature controller instead.

Shut Off Outside Air and Exhaust System

6. Shut off outside air completely when the building is unoccupied and outside air is not needed. Use a separate timer or an EMS unit for this.

7. Simultaneously, when the outside air is closed off, shut off unnecessary exhaust systems such as for toilets, kitchens, conference rooms, labs etc. Use a separate timer or and EMS unit.

Use Economizer For Maximum Free Cooling

8. Use maximum amount of outside air for free cooling when cooling is needed with economizer cycle. If the OA temperature drops somewhere below the return air temperature the OA damper is to open up. Mechanical refrigeration should only go on if the mixed air temperature cannot be held.

9. Must have automatic changeover control in economizer cycle to switch back and forth between minimum outside air damper setting and the settings for using outside air as cooling.

Shut off Cooling and HVAC Units

10. Turn off mechanical cooling system during unoccupied hours.

11. Shut off entire HVAC system where possible when occupied during summer and mild weather and in winter where freeze-ups or morning warm-ups are not a problem.

WHAT TO WATCH OUT FOR WHEN REDUCING MINIMUM OUTSIDE AIR

1. Don't reduce outside air volume below direct exhaust air quantities thereby putting the building under a negative pressure and forcing air infiltration.

2. Maintain some building pressurization. This means bring 1 to 5 percent more outside air than is exhausted.

3. Don't reduce outside air volume below what is needed for healthy ventilation and code requirements.

4. When using OA for cooling at night there are periods when air at higher wet bulb temps will cause condensations in spaces which were cooled during the day.

   Maintaining a higher space temp of 78°F will minimize the amount of time this could occur.

PROBLEM OF COMMON MIXED AIR PLENUM

If systems in common mixed air plenums have different requirements for OA and RA, there can be excessive heating and cooling costs because the OA is generally controlled by the system or zone needing the greatest cooling.

Hence, the systems not needing OA for cooling while one or more of the others do, forces unnecessary heating of the OA in the satisfied systems.

For example: Take a building with four supply systems with a common OA plenum with each HVAC unit requiring minimum outside air of 15 percent:

S-1	30,000 Cfm,	OA 4500 Cfm
S-2	30,000 Cfm,	OA 4500 Cfm
S-3	20,000 Cfm,	OA 3000 Cfm
S-4	20,000 Cfm,	OA 3000 Cfm

If system S-1 needs maximum cooling at a particular point in time, while the other systems are satisfied cooling wise, the common OA damper opens up to meet S-1's cooling need.

However, since the other three systems don't need cool air the excess outside air going through them has to be heated unnecessarily to bring them up to a minimum discharge temperature.

Solution is to put separate OA dampers on intake of each unit. The outside air plenum itself can be maintained as is.

SETTING OUTSIDE AIR DAMPERS

If manual OA:
1.   Read Cfm's of OA, RA, EXA and set dampers.
2.   Set by temperature of mixed air.

If automatic dampers or economizers:
1.   Set OA and RA dampers to be diagrammatically opposite; OA damper 100 percent open when RA damper 100 percent closed and visa versa.

2.   OA damper to open in a certain OA temperature range

3.   Mixed air temperature controller to modulate to maintain fixed mixed air temperature.

4.   Outside air damper to go to minimum when heating or cooling units go on.

Figure 10-7
Analysis of Using Outside Air For Free Cooling

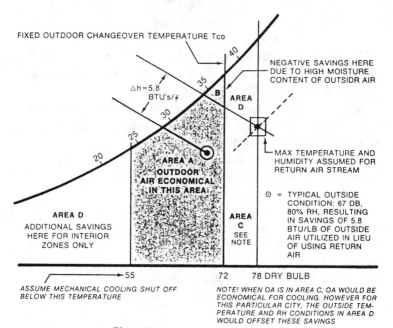

Fixed Temperature Changeover

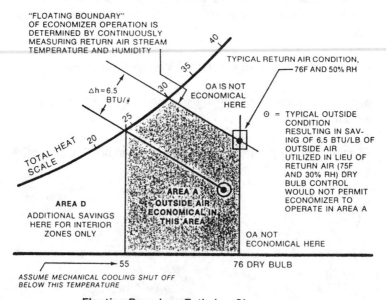

Floating Boundary Enthalpy Changeover

Table 10-1
Weather Data and Economizer Savings

| | | Design Conditions | | | | | Economizer Savings | |
| | | Winter | | Summer | | | | |
City	State	Avg DB Winter Temp	Length In Weeks	Avg DB Summer Temp	Length In Weeks	10^6 Btu To Cool Air	Optimum Changeover Temp For Dry Bulb Economizer	10^6 Btu Cooling Savings
Atlanta	GA	41.1	19.8	78.7	30.0	62.733	72	16.727
Boston	MA	35.1	31.1	76.0	19.8	32.537	72	14.644
Charleston	SC	43.3	14.2	78.7	36.0	90.941	72	15.437
Chicago	IL	34.2	30.0	77.0	20.9	38.791	72	12.434
Cleveland	OH	34.0	29.4	76.5	21.0	34.066	72	12.640
Dallas	TX	42.5	15.1	82.8	34.6	74.244	72	15.239
Denver	CO	35.2	29.4	77.9	22.6	27.052	82	19.493
Detroit	MI	33.8	30.5	75.8	19.2	32.997	72	12.892
Houston	TX	47.0	6.0	80.3	42.0	111.808	67	10.700
Indianapolis	IN	35.8	26.7	78.0	23.9	45.515	72	12.548
Las Vegas	NV	43.7	15.6	86.8	35.4	60.567	82	20.457
Los Angeles	CA	50.2	8.9	72.0	32.6	42.111	72	37.315
Miami	FL	49.3	1.6	80.4	50.1	143.912	67	4.011

(Continued)

Table 10-1 (Continued)

City	State	Design Conditions					Economizer Savings	
		Winter		Summer				
		Avg DB Winter Temp	Length In Weeks	Avg DB Summer Temp	Length In Weeks	10^6 Btu To Cool Air	Optimum Changeover Temp For Dry Bulb Economizer	10^6 Btu Cooling Savings
Milwaukee	WI	33.0	30.0	77.0	20.9	38.700	72	12.400
Minneapolis	MN	29.3	31.0	76.8	18.8	30.661	72	12.968
Nashville	TN	39.3	23.3	79.7	28.4	62.17	72	13.100
New Orleans	LA	46.1	9.4	79.8	39.6	111.48	67	11.137
New York City	NY	38.0	27.5	76.0	20.0	40.689	72	14.825
Philadelphia	PA	38.2	26.0	77.5	23.7	40.689	72	14.161
Portland	OR	44.0	30.8	73.5	15.8	18.097	82	25.334
Richmond	VA	40.9	20.9	77.8	26.8	60.639	72	15.122
Salt Lake City	UT	36.5	30.2	79.0	19.9	25.170	82	15.364
Seattle	WA	43.7	37.3	70.9	9.4	15.768	77	25.142
St. Louis	MO	36.1	24.2	79.6	26.3	52.665	72	12.934
Trenton	NJ	37.5	26.9	77.1	22.9	40.689	72	13.772

Note: Btu based on 10 hour days, 5 day weeks, 1000 Cfm to 55°F for cooling season

METHODS OF OUTSIDE AIR CONTROL

1. **High Limit Constant Mixed Air Temp** Based on mixed air temperature and mixed air controller. Not an accurate method.

2. **Low Limit Control** on leaving side of heating coil. A much more accurate method.

3. **Mixed Air Reset** A constant minimum outside air volume is maintained when outside air is not used for cooling.

4. **Volumetric Put Outside Air Measuring Station** Air measuring station constantly measures outside air volume and automatically controls minimum flow and economizer flow.

5. **Manual Minimum Outside Air Control** A fixed minimum amount of outside air is maintained at all times when the system is in operation.

6. **Two Position Minimum OUTSIDE Air Control** Outside air damper opens up automatically to fixed minimum position when system is in operation and closes when system is turned off.

How to Control Outside Air used for Cooling for Different Type Systems

A. **Multizone Systems in General** When OA is used for cooling, control from space requiring greatest cooling.

B. **Reheat and Induction Systems** Temperature control is maintained by reheating mixed air.

C. **Dual Duct Systems** Temp control is maintained by mixing mixed air and warm air.

D. **Variable Air Volume Systems** Vary volume of cooled mixed air to maintain control of temp.

EVALUATING INDOOR AND OUTDOOR AIR QUALITY

Energy conservation measures that restrict outside air intake can contribute to health problems among office workers. Symptoms of "tight building syndrome" can include headaches, nausea, eye and skin irritation, drowsiness and fatigue.

Indoor pollutants can arise from:

> Too little combustion air in boilers leading to the production of CO instead of CO_2, outgassing.
>
> The emission of formaldehyde from furnishings, curtains, wall coverings or copier machines.
>
> Cigarette smoking; Micro-organisms, especially in humid areas
>
> Contamination of building supply air
>
> Intake of exhaust air in the supply air duct.

1. For each person in the building, at least 20 cubic feet per minute (Cfm) of fresh air should be brought into-the building. Perimeter areas normally require more fresh air, and air flow should be 30 Cfm per person.

2. Make sure the CO_2 level in the building is less than 2,500 parts per million, at a minimum. Problems have been reported in buildings when the CO_2 level exceeds 900 ppm. The CO_2 level in clean outside air is 320 to 330 ppm.

3. Contamination of supply air by exhaust air can be a serious problem in many buildings. ASHRAE recommends placing the supply air intake in the lower third of the building, and the return air intake in the upper third. In many buildings, this may prove impractical; consider increasing the amount of fresh air in the building to compensate for any contamination of intake air.

Table 10-2
Night Setback Energy Savings
Approximate Percent Fuel Savings with Night
and Dual Setback for 25 Cities

	Percent Savings*			
	Night Setback		Dual Setback (Night + Day)	
City	Setback	Setback	Setback	Setback
Atlanta	11	15	20	27
Boston	7	11	15	22
Buffalo	6	10	13	20
Chicago	7	11	14	21
Cincinnati	8	12	17	24
Cleveland	7	10	14	21
Dallas	11	15	20	28
Denver		11	15	22
Des Moines	7	11	14	20
Detroit	7	11	14	21
Kansas City	8	12	16	23
Los Angeles	12	16	22	30
Louisville	9	13	17	24
Milwaukee	6	10	12	19
Minneapolis	5	9	11	18
New York City	8	12	16	23
Omaha	7	11	13	20
Philadelphia	8	12	17	24
Pittsburgh	7	11	15	22
Portland	9	13	17	24
St. Louis	8	12	16	23
Salt Lake City	7	11	14	21
San Francisco	10	14	19	26
Seattle	8	12	17	24
Washington, D.C.	9	13	18	25

*Savings based on an 8-hour setback period during the night and an 8-hour setback during night and day for dual setback for 7 days per week.
©HONEYWELL, INC. 1977

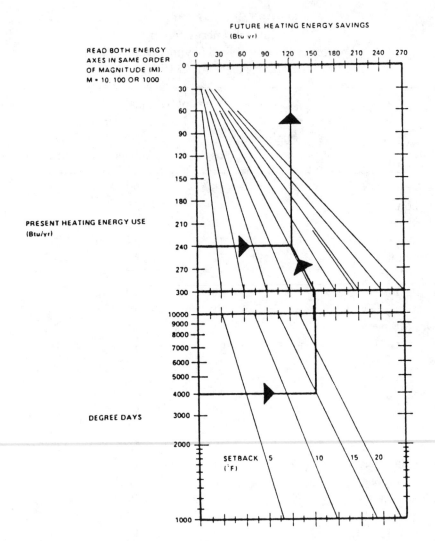

Section 6

Reduce Flows and Resistance

Chapter 11—

Reducing Fan Flows and
Air System Resistance

Chapter 12—

Reducing Pump Flows and
Piping Resistance

Chapter 11

Reducing Fan Flows and Air System Resistance

Reducing fan flows and air system resistance can have a major impact on the break horsepower of fan motors and the consequent electrical energy costs.

Break horsepower varies directly as the cube of the Cfm change and square of the system resistance change. As indicated in chapter 5 on budget operating costs one extra BHP can cost $700 per year running full time and $350 per year running half time at 8 cents per kW. An extra half an inch of static pressure resistance in a system can add a full BHP on the motor.

This chapter will cover the basic elements in evaluating fan performance, using fan charts and curves, applying the fan law, procedure for reducing fan Cfm's, changing drives, package costs, problems and provides a list how to reduce resistance in air systems.

FAN PERFORMANCE

Fan manufacturers publish performance charts for the various sizes and types of fans they produce based on tests and curves. The charts are used for selection of fans and trouble shooting problems in balancing, performance, etc.

The charts are based on the following information:

- Cfm
- Outlet Velocity
- Static Pressure
- RPM
- Break Horsepower

There are three methods of figuring out fan Cfm changes:

1. **Using Fan Curve**
    Using the fan curve is the simplest of all. You simply plot the actual operating point and move up the system resistance curve to the new Cfm required and read off what the new SP, RPM and break horsepowers are.

2. **Using Fan Chart and Calculations**
    The second method is used if a fan curve is not available but a chart is. What you must remember here is that the static pressure always changes when the RPM is changed with a fixed fan and system. The new SP is a function of the new Cfm and RPM and they must be determined first.

3. **Calculations Using Fan Law No. 1**
    In this procedure the new fan performance is calculated using fan law no. 1 formulas.

PROCEDURE FOR REDUCING FAN CFM'S
AND CALCULATING COSTS AND SAVINGS

1. Calculate new peak heating and cooling loads based on energy consumption measures previously implemented or to be implemented and other possible system or building changes from design.

2. Determine new, lower Cfm required based on new heating and cooling loads or other changes.

3. Before taking existing readings on fan:
    *   Make sure filters, coils, etc. are reasonably clean.

    *   Check system for imbalance. If excessive open all outlets 100 percent.

    *   Make sure automatic dampers are operating properly.

    *   Check for duct leakage. Correct if excessive.

    *   Make sure that the system is open and that no fire and manual

dampers are shut.

4. Take readings of actual existing fan performance.
    - Total Cfm Flow
    - Suction and Discharge Fan Static Pressure
    - Fan RPM
    - Filter, Coil, Damper Pressure Drops
    - Outside Air Flow and Operation
    - Running Amps if In Question
    - Record Motor Name Plate Data
    - Check Starter and Overload Size

5. Determine operational hours of fan.

6. If possible put amp recorder on fan for 24 hours on a week day and weekend day.

7. Check the intake and discharge water temperatures, pressures and flows at coils.

    If water flows and temperatures are excessively higher
    or lower than design, determine effect on air flow.

8. If working with DX coils, check if condenser system has unloaders, hot gas by-pass or multiple compressors to avoid freeze up at coils with lower air volumes.

9. Calculate existing annual electrical consumption of fan motor in terms of kWh and costs.

$$kWh = \frac{1.73 \times Amps \times Volts \times Hr}{1000}$$

$$= \frac{1.73 \times 77 \times 460 \times 5260}{1000}$$

$$= 322{,}070$$

$$Old\ Costs = kWh \times Costs/kWh$$

$$= 322{,}070 \times \$.07$$

$$= \$22{,}545$$

10.   Calculate New Fan Performance

System S-1, Suburban Office Building

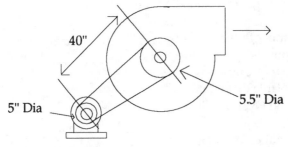

Actual
38,000 Cfm
1,585   Rpm
8"   Sp
77   Amps
460   Volts

Required
26,000 Cfm

Motor Name Plate:
75 Hp
1750 Rpm
460v/3/60Cy
96 Amps Max

$$\text{RPM new} = \text{RPM old}\,\frac{\text{Cfm new}}{\text{Cfm old}} = 1.585 \times \frac{26{,}000}{38{,}000} = 1110\ \text{RPM}$$

$$\text{SP new} = \text{SP old}\,\frac{\text{Cfm new}^2}{\text{Cfm old}} = 8'' \times \frac{26{,}000}{38{,}000} = 4.1''\ \text{SP}$$

$$\text{Bhp actual} = (\text{HP}) \times \frac{\text{Amps act}}{\text{Amps rated}} \times \frac{\text{Volts act}}{\text{Volts rated}}\,\frac{77}{96} \times \frac{460V}{460V} = 59\ \text{Bhp}$$

$$\text{Bhp new} = \text{Bhp old} \times \frac{\text{Cfm new}^3}{\text{Cfm old}} = 59 \times \frac{26{,}000^3}{38{,}000} = 19\ \text{Bhp}$$

If new amp draw falls below minimum running amps, which is 40 percent of the full load amps of the motor, it is more energy efficient to replace the motor with a smaller one that allows a lower amp draw. Also, a more energy efficient motor can be selected at that time.

11. Calculate New Drives
        New pulley size required at new RPM
        New belt length needed

12. Check Electrical System

    Check if new motor, wiring, starter or overloads are needed.

13. Calculate new electrical consumption and costs per year.

$$\text{New Costs} = \frac{1/73 \times 27 \times 460 \times 5260}{1000}$$

$$= 113,000$$

$$\text{New Costs} = 113,000 \times \$.07$$

$$= \$7,911$$

$$\text{Savings} = \$22,545 - \$7,911 = \$14,634 \text{ per year}$$

14. Calculate retrofit costs involved for changing the fan Cfm.

    Inspection and Tests
    Calculations
    New Drives if Required
    New Motor if Required
    Removal, Installation, Adjustments
    Startup and Running Tests

15. Determine payback and return on investment.

16. If change is feasible and worthwhile, proceed with drive and motor changes as required.

17. Perform proper startup, check out performance and monitor.

Table 11-1
Airfoil Fan
Double Width, Double Inlet

INTAKE AREA = 12.98 SQ. FT. • WHEEL DIAMETER 33"

OUTLET AREA = 11.27 SQ. FT. • TIP SPEED F.P.M. = 8.65 × R.P.M. • MAXIMUM B.H.P. = 16.563 $\left(\frac{R.P.M.}{1000}\right)^3$

Top white area, Class I / Top grey area, Class II / Bottom white area, Class III / Bottom grey area, Class IV

C.F.M.	O.V.	1/4" S.P.		3/8" S.P.		1/2" S.P.		5/8" S.P.		3/4" S.P.		7/8" S.P.		1" S.P.		1-1/4" S.P.		1-1/2" S.P.	
		RPM	BHP	RPM	BHP	RPM	BHP	RPM	BHP	RPM	BHP	RPM	BHP	RPM	BHP	RPM	BHP	RPM	BHP
9016	800	321	0.55	355	0.73	387	0.93	418	1.14	447	1.36	476	1.60	503	1.84				
10143	900	343	0.67	376	0.88	405	1.07	434	1.31	461	1.55	488	1.79	514	2.04	563	2.58		
11270	1000	368	0.81	398	1.04	426	1.27	452	1.51	478	1.76	503	2.02	527	2.28	574	2.84	619	3.44
12397	1100	393	0.97	421	1.22	447	1.47	472	1.73	496	2.00	520	2.27	542	2.55	587	3.13	629	3.75
13524	1200	425	1.16	445	1.47	470	1.70	493	1.98	516	2.26	538	2.55	560	2.85	602	3.46	642	4.10
14651	1300	445	1.38	470	1.67	493	1.96	516	2.26	537	2.56	558	2.86	579	3.18	619	3.82	657	4.46
15778	1400	472	1.60	495	1.93	518	2.24	539	2.56	559	2.88	579	3.21	599	3.54	637	4.21	673	4.91
16905	1500	499	1.90	521	2.23	542	2.56	563	2.90	582	3.24	602	3.58	620	3.93	656	4.65	691	5.38
18032	1600	526	2.21	548	2.56	568	2.91	587	3.27	606	3.63	624	3.99	642	4.36	677	5.11	711	5.88
19159	1700	552	2.55	574	2.92	594	3.28	612	3.68	630	4.06	648	4.43	665	4.83	699	5.62	731	6.42
20286	1800	582	2.94	601	3.31	620	3.72	638	4.12	655	4.53	672	4.93	689	5.33	721	6.16	752	7.00
21413	1900	610	3.36	628	3.77	646	4.19	664	4.61	680	5.03	697	5.45	713	5.88	746	6.75	774	7.62
22540	2000	638	3.83	656	4.26	673	4.69	690	5.13	706	5.57	722	6.02	737	6.48	767	7.37	796	8.29
24794	2200	695	4.90	712	5.37	727	5.84	743	6.32	758	6.81	773	7.29	789	7.78	815	8.77	846	9.76
27048	2400	752	6.17	768	6.68	783	7.18	797	7.71	811	8.24	825	8.76	839	9.29	865	10.36	891	11.44
29302	2600	810	7.65	824	8.20	838	8.76	852	9.32	865	9.98	878	10.45	891	11.02	916	12.17	940	13.33
31556	2800	868	9.36	882	9.96	895	10.55	907	11.15	920	11.76	932	12.37	944	12.98	968	14.21	991	15.85
33810	3000	927	11.33	939	11.96	951	12.60	964	13.24	975	13.89	987	14.53	999	15.19	1021	16.50	1043	17.82
36064	3200	985	13.57	997	14.24	1009	14.91	1020	15.60	1031	16.28	1043	16.97	1053	17.66	1075	19.00	1096	20.46
38318	3400	1044	16.09	1055	16.80	1066	17.52	1077	18.24	1088	18.97	1098	19.70	1109	20.43	1129	21.90	1149	23.39

C.F.M.	O.V.	2" S.P. RPM	2" BHP	2-1/2" RPM	2-1/2" BHP	3" RPM	3" BHP	3-1/2" RPM	3-1/2" BHP	4" RPM	4" BHP	4-1/2" RPM	4-1/2" BHP	5" RPM	5" BHP	5-1/2" RPM	5-1/2" BHP	6" RPM	6" BHP
13524	1200	719	5.47	791	6.98														
14651	1300	730	5.91	800	7.44	866	9.10												
15778	1400	743	6.38	810	7.96	874	9.64	935	11.44										
16905	1500	758	6.90	822	8.52	883	10.24	943	12.07	1000	14.00								
18032	1600	775	7.47	836	9.14	895	10.90	952	12.76	1007	14.72	1061	16.77						
19159	1700	792	8.07	852	9.80	908	11.62	963	13.51	1017	15.50	1069	17.59	1120	19.76				
20286	1800	811	8.72	868	10.52	923	12.38	976	14.33	1028	16.36	1078	18.47	1127	20.67	1175	22.97		
21413	1900	831	9.42	886	11.28	938	13.20	990	15.20	1040	17.37	1089	19.43	1137	21.66	1184	23.98	1229	26.38
22540	2000	852	10.16	905	12.09	956	14.08	1006	16.13	1055	18.26	1102	20.45	1148	22.72	1193	25.07	1238	27.50
24794	2200	895	11.79	945	13.86	993	15.98	1040	18.16	1085	20.40	1130	22.69	1174	25.06	1216	27.40	1258	29.99
27048	2400	940	13.62	987	15.84	1033	18.11	1077	20.42	1120	22.78	1162	25.20	1204	27.67	1244	30.20	1288	32.72
29302	2600	987	15.67	1032	18.05	1075	20.46	1117	22.92	1162	25.63	1198	27.96	1238	30.56	1276	33.20	1314	35.90
31556	2800	1036	17.95	1078	20.49	1119	23.06	1159	25.67	1199	28.31	1237	30.99	1274	33.16	1311	36.49	1347	39.30
33810	3000	1085	20.49	1126	23.19	1165	25.91	1204	28.68	1241	31.47	1278	34.30	1313	37.16	1349	40.07	1383	43.01
36064	3200	1136	23.29	1175	26.19	1213	28.95	1249	31.96	1285	34.91	1320	37.89	1355	40.90	1388	43.94	1422	47.02
38318	3400	1188	26.38	1225	29.40	1261	32.45	1296	35.53	1331	38.64	1365	41.77	1398	44.93	1430	48.13	1462	51.35
40572	3600	1240	29.77	1276	32.96	1311	36.17	1345	39.41	1378	42.68	1410	45.97	1442	49.28	1473	52.63	1504	56.00
42826	3800	1293	33.49	1328	36.84	1361	40.22	1394	43.04	1426	47.04	1457	50.49	1488	53.96	1518	57.47	1548	60.99
45080	4000	1347	37.55	1380	41.06	1412	44.60	1444	48.16	1475	51.75	1505	55.36	1535	58.99	1564	62.65	1592	66.33

C.F.M.	O.V.	6-1/2" RPM	6-1/2" BHP	7" RPM	7" BHP	7-1/2" RPM	7-1/2" BHP	8" RPM	8" BHP	9" RPM	9" BHP	10" RPM	10" BHP	11" RPM	11" BHP	12" RPM	12" BHP	13" RPM	13" BHP
24794	2200	1300	32.56	1340	35.25	1381	37.92	1420	40.70	1511	49.67	1583	55.83	1667	66.03	1732	72.79	1810	84.21
27048	2400	1324	35.44	1362	38.16	1400	40.32	1438	43.79	1554	53.24	1600	59.53	1685	70.28	1748	77.13		
29302	2600	1363	38.47	1388	43.08	1425	45.08	1487	51.06	1554	57.24	1620	63.25						
31556	2800	1387	42.57	1418	49.03	1459	52.10	1517	55.23	1582	61.60	1645	68.18	1707	74.96	1767	81.95	1827	89.14
33810	3000	1417	46.00	1457	53.30	1484	56.30	1550	59.73	1645	66.33	1677	73.13	1732	80.07	1791	87.22	1848	94.55
36064	3200	1454	50.24	1494	57.30	1519	61.35	1585	64.99	1681	71.89	1707	78.13	1761	85.59	1817	92.92	1873	100.43
38318	3400	1493	54.61	1525	62.84	1594	66.30	1622	71.89	1681	76.89	1737	84.13	1795	91.52	1847	99.06	1900	106.76
40572	3600	1534	59.40	1564	68.12	1634	71.73	1663	75.36	1718	82.73	1772	90.22	1826	97.85	1878	105.62	1930	113.54
42826	3800	1577	64.54	1606	73.76	1676	77.52	1704	81.20	1758	88.95	1810	96.71	1862	104.60	1913	112.74	1963	120.75
45080	4000	1621	70.03	1648	79.42	1720	83.04	1746	87.52	1798	95.55	1849	103.69	1899	111.79	1949	120.10	1997	128.11
47334	4200	1666	75.84	1693	86.17	1764	90.23	1790	94.32	1840	102.55	1890	110.69	1938	119.32	1986	127.86	2033	136.51
49588	4400	1712	82.14	1738	94.32	1810	97.16	1835	101.43	1884	109.97	1932	118.60	1979	127.12	2025	136.14	2071	145.06
51842	4600	1759	88.78	1785	92.97	1856	104.56	1880	108.96	1928	117.09	2075	126.74	2061	135.76	2045	150.87	2111	153.07
54096	4800	1806	95.84	1832	100.14	1904	112.58	1927	116.93	1972	124.69	2061	135.73	2061	145.65	2108	160.05	2151	162.54
56350	5000	1856	101.32	1880	107.84	1952	120.84	1975	125.35	2020	134.83	2108	145.38	2108	154.00	2151	163.71	2193	173.19
58604	5200	1905	107.26																

Performance shown is for air foil fan with outlet duct. BHP does not include drive loss.
Underlined figures indicate maximum static efficiency.

By reducing the fan flow from 38,000 to 26,000 CFM, the static pressure resistance in the system reduces from 8 inches to 4 inches and the BHP from around 60 to 20.

Chart 11-1
Sample Fan Curve
Reducing Air Flow

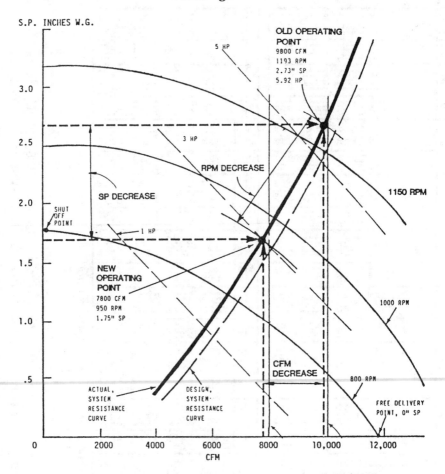

*Fan performance curve showing how operating
point moves on system resistance line.*

The actual curve and design system curve rarely turn out to be
exactly the same. System curves show how the Cfm, SP and HP vary at
different rpms with a fixed air distribution system.

CHANGING DRIVES

When changing drives to alter the fan rpm and volume, certain information and calculations are needed:

1. New RPM ratio

$$New\ RPM\ ratio = Pulley\ Diameter\ ratio$$

$$RPM\ ratio = \frac{RPM\ motor}{RPM\ fan}$$

$$Pulley\ Diameter\ ratio = \frac{DIA\ fan}{DIA\ motor}$$

2. Diameter of sheave not being changed
3. Number of existing belts, type and lengths
4. Shaft and keyway size for bore and keyway sizing on sheave to be changed
5. Center to center distance of shafts
6. Amount of motor movement on motor slide rail to adjust for belt length and tension
7. Motor frame number

The procedure for selecting a new motor or fan sheave and new belts is shown in the following example:

$$DIA\ fan\ sheave = DIA\ motor\ sheave \left(\frac{RPM\ motor}{RPM\ fan}\right) = 5 \times \frac{1725}{1193} = 7.23"\ DIA$$

$$Belt\ Length = 2C + 1.57 \times (D + d) + \frac{(D - d)^2}{4C}$$

$$= 2 \times 40 + [1.57 \times (7.23 + 5)] + \frac{(7.23 - 5)^2}{4 \times 40}$$

$$= 80 + 19.2 + .03 = 99.23\ Inches\ Long$$

RULE OF THUMB ON DRIVE DESIGN

1. Centers should not exceed 2-1/2 to 3 times the sum of the sheave diameters, nor be less than the diameter of the larger sheave.
2. Diameter ratio should not exceed 5 to 1.
3. The angle of contact on the smaller sheave should not be greater than 120 degree with a perpendicular line off that point on the periphery of the sheave.

Table 11-2
BHP Savings in Fan and Pump
Flow Reductions
Based On Actual Flows and No System Changes

Percent Cfm or GPM Reduction	Ratio New Flow to Old	BHP Reduction		S.P. or Head Reduction	
		Percent	Multiplier	Percent	Multiplier
5%	.95	19%	.86	10%	.90
10%	.90	27%	.73	19%	.81
15%	.85	39%	.61	28%	.72
20%	.80	49%	.51	36%	.64
25%	.75	58%	.42	44%	.56
30%	.70	66%	.34	51%	.49
35%	.65	73%	.27	58%	.42
40%	.60	78%	.22	64%	.36
50%	.50	87%	.13	75%	.25
60%	.40	94%	.06	84%	.16
70%	.30	97%	.03	91%	.09

Notes
1. Multipliers based on ratio squared for SP or head reduction. 2. Multipliers based on cubed BHP reduction. 3. Minimum BHP's or amp draws on motors are generally about 40 percent of full load amps and savings on further fan or pump flow reductions beyond this yield no gains.

If BHP reduction exceeds minimum BHP of motor, it is generally advisable to use a new smaller HP motor.

Where:
    SP   = Static Pressure
    BHP = Break Horsepower
    Cfm   = Cubic Feet per Minute air
    GPM = Gallons Per Minute

Table 11-3. Package Costs of Reducing Fan Cfm's

Includes: Readings, New Drives, New Motors, New Thermal Overloads, Startup

New Cfm	New Hp	Direct Material Costs — Motor Dripproof	Direct Material Costs — Drives Old	Labor — Removal	Labor — Motor	Labor — Drives	Labor — Total	Total Material & Labor — Direct Costs	Total Material & Labor — With 30% O&P
Low Pressure, 1" to 3" SP, 2,000 FPM O.V.									
5,000	3	257	143	3.00	3.50	1.70	12.20	$888	$1,155
10,000	5	289	147	3.20	3.80	1.80	12.80	948	1,233
15,000	7.5	440	150	4.10	4.30	2.00	14.40	1,167	1,517
20,000	10	531	162	4.60	5.50	2.10	16.20	1,340	1,742
25,000	15	712	177	5.20	7.00	2.20	18.40	1,625	2,113
20 891	193	6.00	8.50	2.30	20.80	$1,916	$2,491		30,000
40,000	25	1,055	203	6.70	11.00	2.40	24.10	2,222	2,889
50,000	30	1,237	216	7.40	13.50	2.50	27.40	2,549	3,314
60,000	40	1,554	229	8.00	15.80	2.60	30.40	2,999	3,899
Medium and High Pressure, 2" to 8" SP, 3,500 FPM O.V.									
15,000	15	712	150	4.10	4.30	2.00	14.40	$1,438	$1,870
20,000	20	891	162	4.60	5.50	2.10	16.20	1,700	2,211
25,000	25	1,055	177	5.20	7.00	2.20	18.40	1,969	2,559
30,000	30	1,237	193	6.00	8.50	2.30	20.80	2,262	2,941
40 1,554	203	6.70	11.00	2.40	24.10	$2,721	$3,538		40,000
50,000	50	1,882	216	7.40	13.50	2.50	27.40	3,194	4,152
60,000	60	2,382	229	8.00	15.80	2.60	30.40	3,827	4,975
80,000	75	2,886	255	8.60	17.20	2.80	32.60	4,444	5,778
100,000	100	3,709	279	9.50	18.50	3.00	35.00	5,388	7,005

1. Initial readings 2 hours included
   Startup readings 2 hours included

Direct labor costs are $40.00 per hour.

OPTIONS IN FAN CFM REDUCTION

1. Reduce RPM and Cfm by changing drives.

2. Reduce RPM and Cfm by installing smaller motor and changing drives.

3. Install inlet vane dampers. Test out best effect on system and energy costs:
    Fixed setting year round
    One setting for summer, another for winter
    Modulate based on OA temperature

AREAS OF CAUTION AND POTENTIAL PROBLEMS
In Fan Cfm Reductions

Many of the potential problems that occur in variable air volume systems, when the volume of the fan is reduced, also occur when the flow of a constant volume fan is permanently reduced. Reducing fan air flow can adversely effect the following items.

1. DX (direct expansion) coils.
2. Supply outlets, air patterns.
3. Outside air.
4. Return air.
5. Code ventilation requirements.
6. Humidity.
7. Efficiency of heat transfer through coils.
8. Minimum fan Cfm point must not enter surge area.
9. Pressure on inlets of terminals to operate them properly on medium and high pressure systems.

Yearly Energy Consumed
Centrifugal Fans, Backward Curve Blades

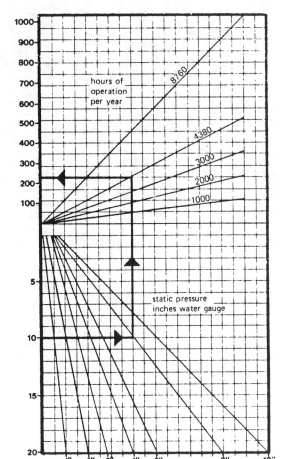

Yearly Energy Consumed-Centrifugal Fans,
Forward Curve Blades

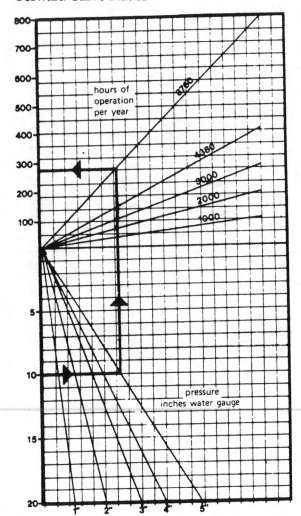

Reducing Fan Flows
Check Off List

Tests
—Retest Fan
—Rebalance System
—Fan Test Readings
Drives
—New Belts
—New Motor Sheave
—New Fan Sheave

Fan
—New Fan
—Inlet Vane Dampers

Electrical
—New Motor
—New Starter Overloads
—Wiring
—New Starter

TROUBLE-SHOOTING FANS

There are many problems with fans that can beset the balance or building engineer.

1. Unbalanced Wheels cause vibrations, noises and premature wearing of bearings and drives. Wheels have to be dynamically balanced and weights put on them for balance.

2. Wheel misalignment and improper wheel overlap and gap with the inlet cone, can cause a sharp loss of air volume and cycling within the fan. Misalignment is when the wheel is not centered on the inlet cone. Improper gaps and overlaps are shown below. Improper overlaps are corrected by moving the wheel on shaft, and centerline misalignment and improper gaps by shifting fan inlet cone.

Figure 11-1.
Improper gaps and overlaps of inlet cone and wheel venturi can cause air cycling and drastic loss in fan volume.

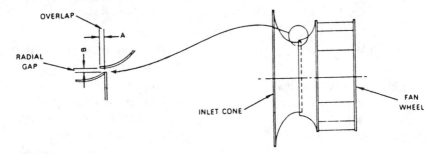

3. Bad fan discharges can cause havoc with air quantities, statics; throbbing, pulsating, noises, duct wreckage. Poor and good discharges are illustrated as follows:

Figure 11-2
Typical effect of discharge connection on velocity pressure

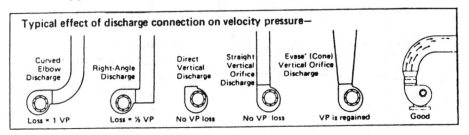

4. A warped shaft, which can be caused by heat or in removing a wheel causes vibrations and premature wearing of bearings and drives.

5. A shaft at an angle in the fan can cause the wheel to rub.

6. Different system curve. The actual system curve may be quite different than shown in the catalogue and air volumes, statics, horsepower may be way different than design.

Reduce Resistance In Air Distribution Systems
Check Off List

Clean Components

— Coils
— Filters
— Clogged Dampers, Louvers Turning Fans, etc.

Lower Face Velocities

— Filters
— Coils
— Through Dampers

Flexible Tubing

— **Straighten** sagging and Winding Tubing and make taut
— Shorten Long flow Tubing Runs

Ductwork Design

— Measure Pressure Drops Across Suspect Fittings
— Install Turning Vanes in Square Elbows where left

Filters
— Clean or replace if dirty
— Use Lower Resistance Filters
— If Filter Efficiency Specified
— Exceeds Filtering requirements use lesser Filter

Sound Attenuators
— Remove if Possible
— Use Acoustical Lining instead
— Install smaller units
— Install units with less Pressure Drop

Coils—
— Clean
— Straighten Out Bent Fins
— Reduce depth of multi-row Cooling Coils
— Lower Face velocity
— Combine Heating and Cooling Coil

Grilles, Diffusers
— Use Low Resistance Outlet Versus Parallel

Dampers
— Clean Clogged Dampers
— Check for Closed Fire Dampers
— Take Fire Damper Frames Out of Air Stream

out
— Beware of Double Offsets, especially Square Type
— Beware of Sharp Change Fittings
— Put Longer Taper on Fan Discharge Fittings
— Improve bad Fan Intake Ductwork
— Improve excessively long Index runs
— Streamline Duct Routing

Reduce Duct Leakage
Seal Connections
— Check for Open Duct Ends

Test and Balance
— Proportionate Balance outlets for minimum Pressure Drop
— Reduce Resistance of Excessively Long Index Runs

Insulate
— Insulate Ducts in Un-conditioned Spaces

Fans
— Reduce Fan Volume to be commensurate with Reduced Heating Loads and Reduced Resistance

Chapter 12

Reducing Pump Flows and Piping Resistance

As with fans, an extra BHP on a pump motor due to excessive volume or due to extra resistance in the system, can cost $700 per year running full time and $350 per year running half time at 8 cents per kWh.

Five extra feet of head of resistance in a system causes a pump to draw an extra BHP.

Many pumps are oversized during the initial design of the system, many are pumping excessive flow because resistances are less than anticipated or there have been changes which reduced the head resistance of the system. Flows from pumps can easily be 10, 20, 30 percent higher than needed.

A pump rated for 500 gpm with a 10 hp motor may be pumping out 20 percent more than needed which results in a BHP increase, when factored into pump laws, of nearly 50 percent. That means the 10 hp increases to 15 and may cost $3,500 more to operate full time at 8 cents per kWh. A negligent, wasteful, unnecessary situation.

In order to test out and change flows on pumps or reduce resistance in a system it is necessary that you have good knowledge of pumps, operations, pump curves, pump laws, various formulas, testing methods, pressure drops of piping and equipment, etc. as contained in this chapter.

TYPES OF PUMPS AND WHEELS

There are two major categories of pumps:

1.   Positive displacement-reciprocating, rotary and screw.

2.   Centrifugal-with a variety of impeller designs classified as plain radial flow, mixed flow and axial flow, each within a volute casing and turbine diffuser type pumps.

Four Types of Centrifugal Impellers

Fig. 12-1 illustrates the two types of centrifugal pumps as well as the four basic types of impellers. The plain impeller has single curvature vanes always curved backwards. Wider impellers have vanes of double curvature with the suction ends twisted. These vanes are called mixed flow Francis type vanes.

It is the plain flow impeller centrifugal pump that is used most frequently in air conditioning and refrigeration. They are used to move chilled, warm, hot and refrigerant condensing water, steam condensate, brine, lubricant oil or refrigerant.

A centrifugal pump is distinguished by a continuous steady flow and characteristic performance curves with a smooth rising head and falling power from maximum capacity to shutoff. The pump presents an easy load for a driver. The starting torque is small and operating load is constant. As a rule a constant speed squirrel-cage induction electric motor, NEMA Design B, with normal starting torque is applicable to drive the pump.

The centrifugal pump is rated on the basis of capacity, volume of liquid per unit time, gallons per minute, (gpm) against a head, feet of water required by the fluid transmission system, and the energy required at a given speed.

There are two major elements in a centrifugal pump assembly—an impeller rotating on a shaft supported in a packed or mechanical seal and bearings, and a casing that is the impeller chamber (volute). The impeller imparts the principal force to the liquid, and the volute guides the liquid from inlet to the outlet, at the same time converting the kinetic, velocity, energy into pressure.

STANDARDS AND CODES

Standards of the Hydraulic Institute (an organization of leading pump manufacturers) define the product, material, process and procedure in the design and testing of any type of pump.

Figure 12-1
Types of centrifugal pumps, four types of impellers

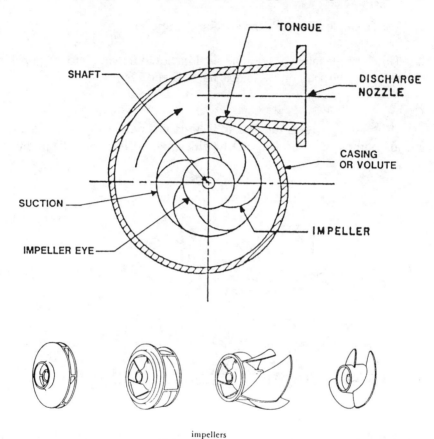

impellers

CLASSIFICATIONS

Impeller

The plain radial flow impeller in a volute casing centrifugal pump is the one usually applied in air conditioning and refrigeration applications.

The impellers are constructed in three arrangements:

1. Enclosed: vanes within an impeller shroud or side walls.
2. Semi-enclosed: vanes assembled with one side wall
3. Open: no side walls, casing serving as side walls

Suction

The liquid approach into the pump may be either:

1.   Thru a single inlet with end suction to impeller.

2.   Thru a single inlet with double suction, liquid flowing into the impel-
     ler along the shaft on two sides, fig. 12-2 and 12-3.

Casing

    The volute casing may be split horizontally, usually with double
suction pumps, or vertically, usually with single suction end inlet pumps.

Figure 12-2.
Single and double suction centrifugal pumps

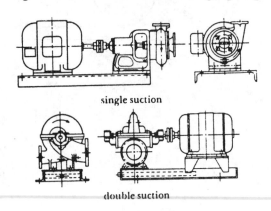

single suction

double suction

Figure 12-3.
Impellers
(Courtesy of Carrier Air Conditioning Co.)

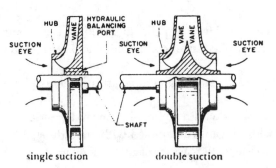

single suction   double suction

Stages

The single stage pump is one with a single impeller; it may have a single or double suction. If the required head is too high for a single impeller to develop, two or more single stage pumps may be used in series, or a set of impellers in series may be put into a single casing. The latter assembly is designated a multi-stage pump.

Rotation

Pump rotation is determined when looking from the drive toward the pump. Thus straddling the drive of a horizontal pump or looking down at the motor end of a vertical pump, if the liquid is entering the suction on the right side (double suction) and moves clockwise within the casing towards the discharge, such a pump is designated as having clockwise rotation. With the suction on the left side and the liquid moving within the casing in counterclockwise direction, the pump is designated as having counterclockwise rotation, fig. 12-4.

Figure 12.4
Rotation designation, centrifugal pump.
(Courtesy of Carrier Air Conditioning Co.)

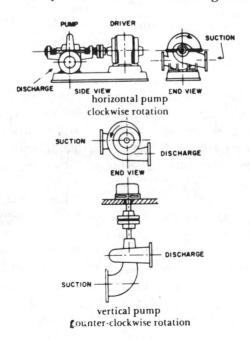

Motors

Of the single-phase motors the capacitor type are used for small pumps. Of the multi-phase motors the standard squirrel-case induction type, NEMA Design B, are the most popular. Occasionally for reasons peculiar either to power distribution regulations or to the customer's economic situation a pump may be driven either by a part-winding, wound rotor or synchronous motor.

CENTRIFUGAL PUMP OPERATION

Basic Theory

The rotating impeller imparts to a fluid a centrifugal force, kinetic energy in the form of velocity. The volute converts about 50 percent of the kinetic energy into the pressure head, potential energy measured in feet of fluid handled. As the fluid flows thru the impeller vanes, a reduced pressure zone is created at the inlet to the vanes.

The atmospheric or system pressure and the static head of the fluid as available act on the pump suction inlet and force the liquid into the pump. This pressure at the pump suction plus the pressure developed by the rotating impeller in the volute produces the flow of the liquid. This is fundamental to the application of the centrifugal pump.

Net Positive Suction Head (NPSH)

Between the pump suction nozzle and the minimum pressure point within the pump impeller, there exists in addition to the suction velocity head, a pressure drop. This pressure drop is due to velocity acceleration, friction and turbulence losses. The suction head, feet of liquid absolute, determined at the suction nozzle and referred to a datum line, less the vapor pressure of the liquid, is called the net positive suction head or NPSH. The suction head necessary to keep liquid flowing into the pump and to overcome the pump internal pressure losses is the required NPSH of the pump.

The required NPSH of a pump is part of the standard design performance data furnished by the manufacturer or of a design specific to a given process pump.

The net positive suction head, in feet of liquid, of the process liquid system, as it exists within the system complex at the entering side of the pump is called the available NPSH. It must be at least equal to or greater than the required NPSH in order to produce a flow thru a pump. A safety factor should be considered to cover a possible excess of required NPSH.

The available NPSH if the algebraic sum determined by the formula:

$$\text{Available NPSH} = \frac{2.31 \, (Pa - PvP)}{Sp \, gr} + Hs - Hf$$

where

NPSH = net positive suction head (absolute pressure, ft)

Pa = atmospheric pressure (absolute pressure, psia) in an open system; or pressure (absolute, psia) within a totally closed system.

Pvp = vapor pressure (psia) of the fluid at pumping temperature; in a totally closed system it is part of the total pressure Pa.

Hs = elevation head, static head (ft) above or below the pump center line. If above, positive statichead; if below, negative static head, sometimes termed suction lift.

Hf = friction head (ft) on the suction side of the system including piping, fitting, valves, heat exchangers at the design velocity (V8 in ft per sec) within suction system.

sp gr = specific gravity of liquid handled at operating temperature.

Suction Lift

The suction lift of the open systems is infrequently a factor in the design of air conditioning and refrigeration systems. Basically a pump does not lift; to operate, it must have pressure at its suction.

To quote from paragraph B-44 of the Hydraulic Institute Standards, "Among the more important factors affecting the operation of a centrifugal pump are the suction conditions. Abnormally high suction lifts (low NPSH) beyond the suction rating of the pump usually cause serious reductions in capacity and efficiency, often leading to serious trouble from vibration and cavitation."

Cavitation

The lack of available NPSH shows up particularly in pump cavitation. If the pressure at any point inside the pump falls below the operating vapor pressure of the fluid, the fluid flashes into a vapor and forms bubbles. These bubbles are carried along in the fluid stream until they reach a region of higher pressure. Within this region the bubbles collapse or implode with a tremendous shock on the adjacent surfaces. Cavitation accompanied by low rumbling or short rattling noise and even vibration causes mechanical destruction in the form of pitting or erosion.

Do not tamper with the pump suction inlet; do not request the pump manufacturer to enlarge the pump suction in order to decrease the required NPSH. The pump efficiency falls off and the whole performance of the impeller is upset.

Vortex

A whirling fluid forming an area of low pressure at the center of a circle is called a vortex. This is caused by a pipe suction placed too close to the surface of the fluid. Such a vortex impairs the performance of a pump and may cause a loss of prime.

In the case of pump suction in a shallow water sump such a vortex may be prevented by placing a plate close to the intake at a distance of one-third diameter from the suction inlet. The plate should extend 2 1/2 diameters in all directions from the center of the inlet.

The performance of a centrifugal pump is affected when handling more viscous fluids. The effects are a marked increase in brake horsepower and in head decrease in capacity and efficiency.

OPERATION IN A SYSTEM

A given centrifugal pump operates along its own head-capacity curve. At full capacity flow the operating point falls at the crossing of the pump head capacity curve and the system head curve, point 1, Fig. 12-5. If the pump is throttled, the operating point moves up the head-capacity curve (Point 2); if it is desired to obtain greater flow to operate down the head-capacity curve (Point 3), the path of flow in the system must be eased to reduce the friction losses. Otherwise the pump must be speeded up or the impeller increased in diameter. Then a new head-capacity curve is established (Point 4). The engineer must carefully analyze the system and select the pump from the manufacturer's performance head-capacity curves.

If the system head is overestimated and the pump is selected with a high head-capacity curve, unfortunate results may follow. The pump will operate on its head-capacity curve to produce an increased flow at decreased head and increased horsepower demand, Fig. 12-6. The system head should always be calculated without undue safety factor extension or as close as practical to the true values to eliminate possible waste of horsepower or possible overload of pump motor with an unvalved system.

Figure 12-5.
Crossover Point of pump and system curves.
(Courtesy of Carrier Air Conditioning Co.)

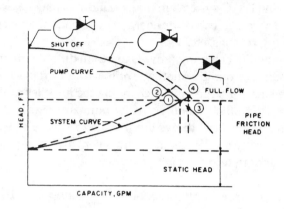

Figure 12-6.
Effect of overestimating pump head.
(Courtesy of Carrier Air Conditioning Co.)

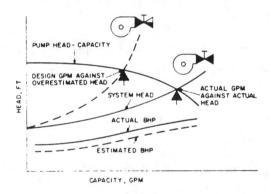

System Head

A flow of liquid within any system of piping including fittings, valves and heat exchangers requires a system head consisting of a velocity head (usually insignificant) and friction head, and must overcome a static head. Thus in any piping system the system head is the algebraic sum of the static head on the pump discharge minus the static head on the pump suction plus the friction losses thru the entire system of fluid flow.

PROCEDURE FOR REDUCING PUMP FLOWS AND
CALCULATING COSTS, SAVINGS AND PAYBACKS

Pumps in HVAC systems in many buildings may be oversized and the flows must be decreased for energy conservation, efficiency, performance, comfort, reasons, etc.

One method of changing flows is to install different diameter impellers in the existing pump. Another approach is to replace the existing pump with a small or larger one, if changing the impeller can't do the job.

In energy retrofits generally smaller pumps at higher efficiency levels and at minimum BHP draws are installed in place of existing ones, rather than inserting smaller diameter impellers in existing pumps, which has its limitations in terms of energy savings.

1. The first step in changing the GPM of a pump is to determine the actual flow of the existing pump and other operating characteristics.

    •   Check and record name plate data on motor and pump.

    •   Read amp draw and voltage of pump motor at full flow at the starter.

    •   If a photo tachometer or stroboscopic light are available read PPM'S of pump.

    •   If a flow measuring device is available in the discharge pipe read the total flow directly from it.

    •   If no flow measuring device is available read the suction and discharge pressures at full flow and determine the GPM flow from the pump curve.

    •   Doing this requires knowing five items of information on the pump.

        Pump Size and Model
        Actual Total Head
        Amps and Volts (to determine the BHP)
        RPM
        Impeller Diameter

- Determine the pump impeller size by slowly closing the balancing cock or gate valve in the discharge line and reading the block tight suction and discharge pressures. Locate the total block tight feet of head on the pump curve at zero flow on the left side of the graph. Note which impeller diameter line intersects or is closest to the feet of shut off head. This is the diameter of the pump.

- Using the existing chilled water pump P-1 in the suburban office building from Chapter 5 as an example, we find the following by inspection and readings:

PUMP: Model 4 E
          1750 RPM
          9 FT HD
          11 " DIA.

MOTOR: 25 HP
          40 AMPS
          460 Volts

- If there is no flow measuring device available, plot this information on the pump curve shown on page 12-10. It can be determined by the intersection of the horizontal feet of head line and the curved impeller diameter line and by reading straight down vertically from there, that:

Actual Existing Flow = 744 GPM.

2. The second step in changing the GPM of a pump is to determine the new GPM required.

- In order to do this the new Btu heat transfer amount must be determined first.

For example, in the suburban office building in chapter 5 the overall Cfm is being reduced 30 percent, from 121,000 to 85,000 Cfm.

New Total Heat Transfer Cooling

Btuh $= 4.5 \times$ Enthalpy Change $\times$ Cfm
        $= 4.5 \times (29 - 23) \times 85,000$
        $= 2,295,000$ Btuh

Figure 12-7.
Pump Characteristic Curve—Existing Pump

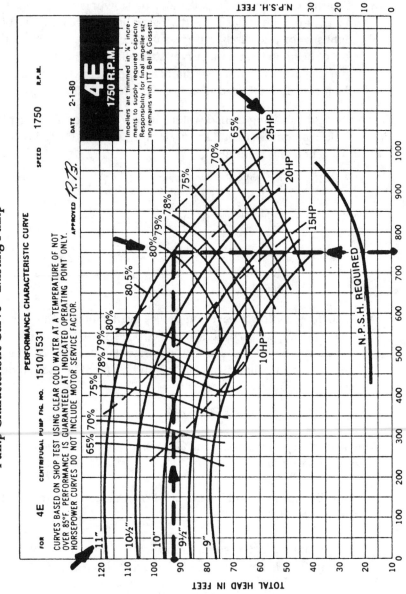

New GPM

$$GPM = \frac{BTUH}{Temp.\ Diff. \times 500}$$

$$= \frac{2,295,000}{10 \times 500}$$

$$= 459\ Gpm$$

3. Now, knowing the actual GPM, ft of head and efficiency, the existing BHP can be calculated.

$$BHP = \frac{GPM \times ft\ head}{3960 \times eff.\ pump}$$

$$= \frac{744 \times 93}{3960 \times .70}$$

$$= 25\ BHP$$

4. Next calculate the new pump impeller diameter, ft of head and BHP using the pump law.

$$IMP\ DIA\ NEW = IMP\ DIA\ old \times \frac{GPM\ new}{GPM\ old}$$

$$= 11 \times \frac{459}{744}$$

$$= 11 \times 62$$

$$= 6.8\ inches$$

$$FT\ HD\ new = FT\ HD\ old \times \frac{GPM\ new^2}{GPM\ old^2}$$

$$= 93 \times (.62)^2$$

$$= 38$$

$$BHP\ new = BHP\ old \times \frac{GPM\ new^3}{GPM\ old^3}$$

$$= 22.5 \times (.62)^3$$

5. By inspection of the fan curve it's apparent that the existing pump cannot effectively achieve the full savings in energy and that a new pump should be installed as follows:

   459 GPM
   38 FT HD
   1750 RPM
   8.0 inch DIA Impeller
   7-1/2 HP Motor
   6.0 BHP

6. The costs of installing 2 new pumps, one for the chiller and one for the condenser, as shown in chapter 8 is $8206.

7. The savings in reducing from 25 BHP to 6.0 BHP on the two pumps, running about 4000 hours per year is:

   2 pumps × 19 BHP = 38 BHP savings
   @ $228/year = $8,664 savings

8. Payback is as follows:

   $$\frac{\$8,200}{\$8,664} = 1.0 \text{ years}$$

CHANGING PUMP GPM'S

Fixed Pump Size and System

   • **Pump Law**

   $$\frac{\text{Impeller}}{\text{DIA. New}} = \frac{\text{Impeller}}{\text{DIA. Old}} \times \frac{\text{GPM new}}{\text{GPM old}}$$

   $$\text{Head new} = \text{Head old} \times \frac{\text{GPM new}^2}{\text{GPM old}}$$

   $$\text{BHP new} = \text{BHP old} \times \frac{\text{GPM new}^3}{\text{GPM old}}$$

Figure 12-8.
Performance Characteristic Curve—New Pump

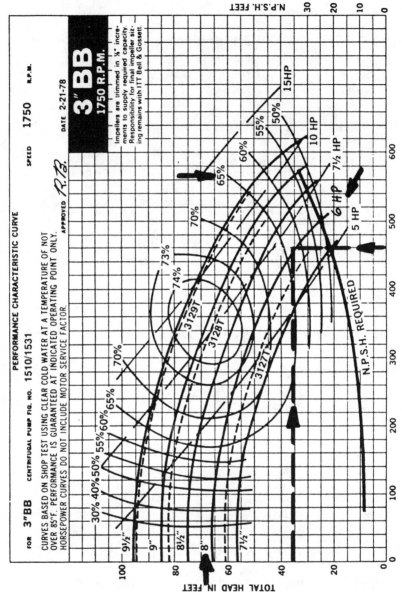

PERFORMANCE CHARACTERISTIC CURVE

for **3"BB** CENTRIFUGAL PUMP FIG. NO. 1510/1531 SPEED 1750 R.P.M.

CURVES BASED ON SHOP TEST USING CLEAR COLD WATER AT A TEMPERATURE OF NOT OVER 85°F. PERFORMANCE IS GUARANTEED AT INDICATED OPERATING POINT ONLY. HORSEPOWER CURVES DO NOT INCLUDE MOTOR SERVICE FACTOR.

APPROVED *R.B.*

3"BB
1750 R.P.M.

Impellers are trimmed in ⅛" increments to supply required capacity. Responsibility for final impeller sizing remains with ITT Bell & Gossett.

DATE 2-21-78

N.P.S.H. FEET

TOTAL HEAD IN FEET

Where:

DIA	=	diameter
GPM	=	gallons per minute
HEAD	=	resistance of system
BHP	=	break horsepower
effM	=	efficiency of motors

- **BHP Formulas (Power Consumption of Water Pump)**

$$BHP = \frac{GPM \times ft\ head}{3960 \times effM}$$

$$\frac{BHP\ actual}{(3\ phase)} = \frac{1.73 \times amps \times volts \times eff \times power\ factor}{746}$$

$$\frac{BHP\ actual}{(rule\ of\ thumb)} = \frac{(name\ plate)}{horse\ powe} \times \frac{amps\ act.}{amps\ rated} \times \frac{volts\ act.}{volts\ rated}$$

USING VALVES FOR FLOW MEASUREMENT

C_V Flow Coefficient Definition
  Gallons flow in one minute
  at 1 PSI pressure drop

Flow thru Valve Definition
  GPM thru valve
  = C_V times sq. root of pressure drop

Formula $GPM = C_v \sqrt{\dfrac{\Delta P}{SpGv}}$

METHODS OF DETERMINING HYDRONIC FLOW

1. Direct Flow Readings with:
   - Venturi Meter
   - Orifice Plate Meter
   - Pitot Tube
   - IL Balometric Valve

2. Pressure Drop Across HVAC Equipment
   • Chiller
   • Terminal Unit

3. Pressure Drop Across Valve Using C_v

4. Pressure Drop Across Pump

5. Thermal Method

Figure 12-9
Pressure Differential Across Pump

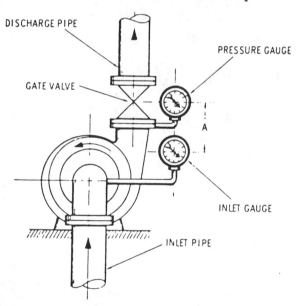

HYDRONIC FORMULAS

• Converting PSI to Feet of Head:

ft hd = 2.31 × psi	inches hd = 27.2 × psi
psi =.433 × ft hd,	psi =.036 × inches hd

• Water Heat Transfer Formulas

$$Btuh = GPM \times (T\ in\text{-}T\ out) \times 500$$

$$GPM = \frac{BTUH}{(T\ in - T\ out) \times 500}$$

- **Electrical Power Consumption of Water Pump**

$$BPH = \frac{GPM \times ft\ head}{3960 \times eff}$$

- **Using System Component As Flow Measuring Device**

$$GPM\ actual = GPM\ design \times \sqrt{\frac{\Delta P\ actual}{\Delta P\ design}}$$

$$\Delta P\ actual = \Delta P\ design \times \left(\frac{GPM\ actual}{GPM\ design}\right)^2$$

- **Coil or Chiller GPM**

$$GPM = \frac{Tons \times 24}{T\ in - T\ out}$$

- **Condensor GPM**

$$GPM = \frac{Tons \times (kW \times /284}{T\ out - T\ in}$$

where:

GPM	=	gallons per minute
ΔP	=	change in pressure across component
Btuh	=	British thermal units per hour
T	=	temperature, °F
BHP	=	break horsepower
kW	=	Kilowatts

Figure 12-9.
Pipe Sizing and Friction Losses

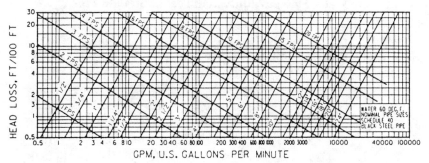

Friction Loss for Water in Commercial Steel Pipe (Schedule 40)

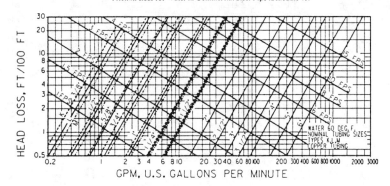

Friction Loss for Water in Copper Tubing (Types K, L, M)

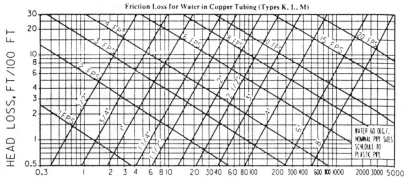

Friction Loss for Water in Plastic Pipe (Schedule 80)

Table 12-1.
Approximate Resistance of Common Fittings
to Water Flow, in Equivalent Feet of Straight Pipe

Nominal Diameter, In.	Standard Ell	Medium Ell	Long Sweep Ell	45° Ell	Gate Valve, Open	Globe Valve, Open
1/2	1.0	0.9	0.7	0.6	0.23	5.0
3/4	1.51	1.3	1.0	0.80	0.40	10.0
1	1.8	1.5	1.3	1.0	0.55	13.0
1-1/4	2.7	2.4	2.1	1.5	0.80	18.0
1-1/2	3.0	2.6	2.3	1.6	0.90	20.0
2	4.1	3.5	3.0	2.3	1.3	29.0
2-1/2	5.5	4.8	3.8	2.8	1.4	35.0
3	7.0	6.0	4.6	3.5	1.9	45.0
4	10.0	8.5	6.6	5.0	2.6	60.0
6	16.0	14.0	12.0	8.0	4.5	110.0
8	21.0	18.0	15.0	12.0	6.0	150.0
10	26.0	22.0	18.0	15.0	8.0	200.0
14	40.0	35.0	31.0	23.0	12.0	280.0
18	55.0	45.0	40.0	30.0	16.0	380.0
20	60.0	55.0	42.0	34.0	17.0	420.0

Table 12-2.
Loss of Head Due to Friction

Gallons Per Min. Delivered	Pipe Sizes, Inches—Inside Diameter					
	1	1-1/4	1-1/2	2	2-1/2	3
20	2.52	0.89	0.42	0.146	0.067	0.038
25	3.84	1.33	0.62	0.218	0.101	0.057
30	5.44	1.88	0.88	0.307	0.142	0.083
35	7.14	2.50	1.18	0.408	0.189	0.107
40	9.12	3.20	1.50	0.528	0.242	0.137
45	—	4.00	1.86	0.656	0.308	0.173
50	—	4.80	2.27	0.792	0.365	0.207
70	—	9.04	4.24	1.430	0.683	0.286
75	—	—	4.80	1.670	0.781	0.458
90	—	—	6.72	2.320	0.991	0.600
100	—	—	8.16	2.860	1.320	0.744
125	—	—	—	4.320	2.060	1.140
150	—	—	—	6.080	2.810	1.580
175	—	—	—	8.160	3.720	2.100
200	—	—	—	10.320	4.740	2.670
250	—	—	—	—	7.260	4.080

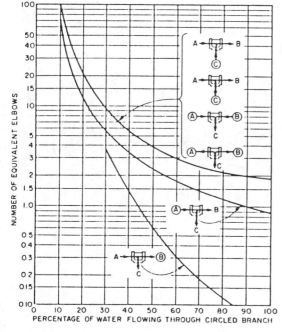

Figure 12-10.
Elbow Equivalents of Tees
at Various Flow Conditions

Table 12-3.
Loss of Head in Feet Due to Friction in Pipes
(per 100 ft. of 15-yr-old iron pipe; for new and smooth iron pipe, use 0.71 of value shown in table

Gallons per Minute	1/2" Pipe		3/4" Pipe		1" Pipe		1-1/4" Pipe		1-1/2" Pipe		2" Pipe		2-1/2" Pipe		3" Pipe	
	Vel.	Fric.	Vel.	Fric.	Vel.	Fric.	Vel.	Fric.	Vel.	Fric.	Vel.	Fric.	Vel.	Fric.	Vel.	Fric.
1	1.05	2.1	—	—	—	—	—	—	—	—	—	—	—	—	—	—
2	2.10	7.4	1.20	1.9	—	—	—	—	—	—	—	—	—	—	—	—
3	3.16	15.8	1.80	4.1	1.12	1.26	0.88	0.57	0.63	0.26	—	—	—	—	—	—
4	4.21	27.0	2.41	7.0	1.49	2.14	1.07	0.84	0.79	0.39	—	—	—	—	—	—
5	5.26	41.0	3.01	10.5	1.86	3.26	—	—	—	—	—	—	—	—	—	—
10	10.52	147.0	6.02	38.0	3.72	11.70	2.14	3.06	1.57	1.43	1.02	0.50	0.85	0.17	0.45	0.07
15	—	—	9.02	88.0	5.60	25.00	3.20	8.60	2.236	3.00	1.53	1.00	0.98	0.36	0.68	0.15
20	—	—	12.03	136.0	7.44	42.00	4.29	11.10	3.15	5.20	2.04	1.82	1.31	0.61	0.91	0.25
25	—	—	—	—	9.30	64.00	5.36	16.60	3.94	7.80	2.55	2.73	1.63	0.92	1.13	0.38
30	—	—	—	—	11.15	89.00	6.43	23.60	4.72	11.00	3.06	3.84	1.96	1.29	1.36	0.54
35	—	—	—	—	13.02	119.00	751	31.20	5.51	14.70	3.57	5.10	2.20	1.72	1.59	0.71
40	—	—	—	—	14.88	162.00	8.58	40.00	6.30	18.80	4.08	6.60	2.61	2.20	1.82	0.91
45	—	—	—	—	—	—	9.65	50.00	7.08	23.20	4.50	8.20	2.94	2.80	2.05	1.15
50	—	—	—	—	—	—	10.72	60.00	7.87	28.40	5.11	9.90	3.27	3.32	2.27	1.38
70	—	—	—	—	—	—	15.01	113.00	11.02	53.00	7.15	18.40	4.58	6.20	3.18	2.67
90	—	—	—	—	—	—	—	—	14.17	84.00	9.19	28.40	5.88	9.80	4.09	4.08
100	—	—	—	—	—	—	—	—	15.74	102.00	10.21	35.80	6.54	12.00	4.54	4.96
120	—	—	—	—	—	—	—	—	18.89	143.00	12.25	50.00	7.84	16.80	5.45	7.00
140	—	—	—	—	—	—	—	—	22.04	199.00	14.30	67.00	9.15	23.30	6.35	9.20
160	—	—	—	—	—	—	—	—	—	—	16.34	86.00	10.46	29.00	7.28	11.80
180	—	—	—	—	—	—	—	—	—	—	18.38	107.00	11,76	36.70	8.17	14.90
200	—	—	—	—	—	—	—	—	—	—	20.42	129.00	13.07	43.10	9.08	17.00
220	—	—	—	—	—	—	—	—	—	—	22.47	154.00	14.38	52.00	9.99	21.30
240	—	—	—	—	—	—	—	—	—	—	24.51	182.00	15.69	61.00	10.89	25.10
260	—	—	—	—	—	—	—	—	—	—	26.55	211.00	16.99	70.00	11.80	19.20
280	—	—	—	—	—	—	—	—	—	—	—	—	18.30	81.00	12.71	33.40
300	—	—	—	—	—	—	—	—	—	—	—	—	19.61	92.00	13.62	38.00

Vel. — Velocity feet per second   Fric. — Friction head in feet

Table 12-4. Loss of Head in Feet Due to Friction in Pipes

(per 100 ft. of 15-yr-old iron pipe; for new and smooth iron pipe, use 0.71 of value shown in table

Gallons per Minute	4" Pipe		5" Pipe		6" Pipe		8" Pipe		10" Pipe		12" Pipe		14" Pipe		15" Pipe		16" Pipe		20" Pipe	
	Vel.	Fric.	Vel.	Fric.	Vel.	Fric.	Vel.	Fric.	Vel.	Fric.	Vel.	Fric.	Vel.	Fric.	Vel.	Fric.	Vel.	Fric.	Vel.	Fric.
40	1.02	0.22	—	—	—	—	—	—	—	—	—	—	—	—	—	—	—	—	—	—
45	1.17	0.28	—	—	—	—	—	—	—	—	—	—	—	—	—	—	—	—	—	—
50	1.28	0.34	—	—	—	—	—	—	—	—	—	—	—	—	—	—	—	—	—	—
70	1.79	0.63	1.14	0.21	—	—	—	—	—	—	—	—	—	—	—	—	—	—	—	—
75	1.92	0.73	1.22	0.24	—	—	—	—	—	—	—	—	—	—	—	—	—	—	—	—
100	2.55	1.23	1.63	0.39	1.14	0.14	—	—	—	—	—	—	—	—	—	—	—	—	—	—
120	3.06	1.71	1.96	0.57	1.42	0.25	—	—	—	—	—	—	—	—	—	—	—	—	—	—
125	3.19	1.86	2.04	0.64	1.48	0.28	—	—	—	—	—	—	—	—	—	—	—	—	—	—
150	3.84	2.55	2.45	0.88	1.71	0.32	—	—	—	—	—	—	—	—	—	—	—	—	—	—
175	4.45	3.36	2.86	1.18	2.00	0.48	—	—	—	—	—	—	—	—	—	—	—	—	—	—
200	5.11	4.37	3.27	1.48	2.28	0.62	—	—	—	—	—	—	—	—	—	—	—	—	—	—
225	6.32	6.61	3.67	1.86	2.57	0.74	—	—	—	—	—	—	—	—	—	—	—	—	—	—
250	6.40	6.72	4.08	2.24	2.80	0.92	1.60	0.22	—	—	—	—	—	—	—	—	—	—	—	—
275	7.03	7.99	4.50	2.72	3.06	1.15	1.73	0.27	—	—	—	—	—	—	—	—	—	—	—	—
300	7.66	9.38	4.90	3.15	3.40	1.29	1.90	0.36	—	—	—	—	—	—	—	—	—	—	—	—
350	8.90	12.32	5.72	4.19	3.98	1.69	2.20	0.41	—	—	—	—	—	—	—	—	—	—	—	—
400	10.20	15.82	6.54	6.33	4.54	2.21	2.60	0.56	—	—	—	—	—	—	—	—	—	—	—	—
450	11.50	19.74	7.35	6.65	5.12	2.74	2.92	0.64	1.80	0.21	—	—	—	—	—	—	—	—	—	—
475	12.30	22.96	7.88	7.22	5.55	3.21	3.10	0.79	1.94	0.25	—	—	—	—	—	—	—	—	—	—
500	12.77	24.08	8.17	812	5.60	3.26	3.20	0.81	2.04	0.28	1.42	0.11	—	—	—	—	—	—	—	—

(Continued)

Table 12-4. Loss of Head in Feet Due to Friction in Pipes (Continued)

(per 100 ft. of 15-yr-old iron pipe; for new and smooth iron pipe, use 0.71 of value shown in table)

Gallons per Minute	4" Pipe		5" Pipe		6" Pipe		8" Pipe		10" Pipe		12" Pipe		14" Pipe		15" Pipe		16" Pipe		20" Pipe	
	Vel.	Fric.	Vel.	Fric.	Vel.	Fric.	Vel.	Fric.	Vel.	Fric.	Vel.	Fric.	Vel.	Fric.	Vel.	Fric.	Vel.	Fric.	Vel.	Fric.
550	—	—	8.99	9.66	6.16	3.93	3.52	0.96	2.25	0.33	1.57	0.14	—	—	—	—	—	—	—	—
600	—	—	9.80	11.34	6.72	4.70	3.84	1.16	2.46	0.39	1.71	0.15	—	—	—	—	—	—	—	—
650	—	—	10.62	13.16	7.28	5.50	4.16	1.34	2.66	0.46	1.85	0.19	1.37	0.09	—	—	—	—	—	—
700	—	—	11.44	15.12	7.84	6.38	4.48	1.54	2.88	0.52	2.00	0.22	1.47	0.10	—	—	—	—	—	—
750	—	—	12.26	17.22	8.50	7.00	4.80	1.74	3.06	0.58	2.13	0.24	1.58	0.11	—	—	—	—	—	—
800	—	—	—	—	9.08	7.90	5.12	1.97	3.28	0.67	2.27	0.27	1.68	0.13	—	—	—	—	—	—
850	—	—	—	—	9.58	8.75	5.48	2.28	3.48	0.75	2.41	0.31	1.79	0.14	—	—	—	—	—	—
900	—	—	—	—	10.30	10.11	5.75	2.46	3.68	0.83	2.58	0.34	1.89	0.16	—	—	—	—	—	—
950	—	—	—	—	10.72	10.71	6.06	2.87	3.88	0.91	2.70	0.35	2.00	0.17	1.73	0.1—	—	—	—	—
1000	—	—	—	—	11.32	12.04	6.40	3.02	4.08	1.01	2.84	0.41	1.20	0.19	1.82	0.14	—	—	—	—
1100	—	—	—	—	12.50	14.31	7.03	3.51	4.50	1.20	3.13	0.49	2.31	0.23	2.00	0.16	—	—	—	—
1200	—	—	—	—	13.52	16.68	7.67	4.26	4.91	1.45	3.41	0.57	2.52	0.28	2.18	0.19	—	—	—	—
1500	—	—	—	—	—	—	9.60	6.27	5.10	2.09	4.20	0.65	3.15	0.39	2.73	0.28	2.39	0.24	—	—
2000	—	—	—	—	—	—	12.70	10.71	8.10	3.50	5.80	1.43	4.20	0.66	3.64	0.47	3.19	0.39	—	—
2500	—	—	—	—	—	—	—	—	10.10	6.33	7.00	2.18	5.25	1.01	4.55	0.72	3.99	0.58	—	—
3000	—	—	—	—	—	—	—	—	12.10	7.42	8.40	3.39	6.30	1.57	5.46	1.12	4.79	0.80	3.08	0.27

Vel. — Velocity feet per second   Fric. — Friction head in feet

REDUCE RESISTANCE OF PIPING SYSTEMS
Check Off List

Clean Components
— Dirty Strainers
— Steam Traps
— Valves

Leakage
— Steam Traps
— Valves

Valve Settings
— Open All Throttling Valves
— Check For Defective Valves

Piping System
— Check For High Resistance Fittings
— General Poor Piping Design
— Higher Velocities Than needed
— Install Larger Piping For Lower Pressure Drops

Steam
— Install PRV's in Steam Lines That Require Lower Loads
— Install Condensate Return Lines Where Condensate Is Wanted

Reduce Flows
— Smaller Impeller
— Smaller Pump

Pumps
— Use 2 Pumps in Parallel. One Handling 1/3 Load For Part Loading and the Other 2/3 Load
— Go to Variable Volume System

REDUCING PUMP FLOW

Tests
— Pump Test Readings
— Retest Pump
— Rebalance System

Pump
— New Impeller
— New Motor
— New Pump

Valves & Piping
— Gate Valves
— Flow Measuring Device
— Readout Stations
— Unions
— Fittings, Pipe

Electrical
— New Starter Overloads
— Wiring
— New Starter

Test and Balance
— Check Pressure Drops
   Across:
   — Steam Traps
   — Pumps
   — Strainers
   — Boilers
   — Chillers
   — Coils
   — Cooling Tower

— Total Pump Flow
— Check Flows Through Main Lines

Section 7

HVAC System Conversions

Chapter 13 -

Variable Air Volume Conversions

Chapter 13

Variable Air Volume Conversions

There are many potential areas where energy can be saved in variable air volume systems over constant air volume operations such as with the fans, pumps, cooling and heating equipment eliminating wasteful reheating and recooling and consequent savings in heating and cooling.

The procedure for evaluating and implementing a VAV conversion is quite extensive, but very important. Knowledge of the different types of VAV systems, VAV terminal units, control systems as well as methods of varying fan volume is essential.

POTENTIAL AREAS OF VARIABLE AIR VOLUME SAVINGS

There are many potential areas where energy can be saved in variable air volume systems over constant volume operation.

A. Fan Savings
1. Fan electrical energy increases with fan volume according to the cube fan law.

2. On an average during VAV operation fans run at 40 percent less than full load during occupancy operation.

3. Fans in the past for HVAC systems were selected based on calculated total worst possible load concepts, which may on an average, be 25 percent too high.

These peak calculated loads are generally too high for various reasons. All the parts of the building generally don't reach their peak loads simultaneously and hence will never need a fan running at that capacity. The calculated loads may carry design fudge factors of 5 to 20 percent, and various energy conservation methods in recent years may have reduced the load needed on the fan.

Constant volume systems run constantly at these excessively high air volume loads and waste a great deal of energy. Whereas, a supply fan in a VAV system running at 38,000 Cfm will be automatically reduced possibly 30 percent to a 26,000 Cfm actual peak load requirement.

4. When fans must run during unoccupied hours during nights and weekends due to weather conditions, piping freezing, skin loads etc., they can automatically run at their minimum capacities with VAV systems. This could well be a turndown of 75 percent of full load.

B. **Cooling and Heating Equipment Savings**
    1. If less air passes over cooling and heating coils, less energy is needed to generate the cooling or heating.

    2. Cooling and heating equipment can cycle on and off, be completely turned off for periods of time or cycle through stages commensurate with VAV variations in loads and generate large electrical savings.

    3. The efficiency of the cooling and heating equipment can be increased and smaller equipment may be used with VAV systems taking advantage of diversity.

C. **Reheat and Dual Duct VAV Conversion Savings**
    1. Variable air volume systems eliminate wasteful reheating and recooling of air already cooled or heated.

    There can be a savings of about 30 percent in heating costs in the frost belt states.

    Cooling costs savings can be in the area of 5 percent in the frost belt states and about 10 percent in the sun belt states by eliminating constant volume terminal reheat and dual duct systems.

    In terminal reheat and dual duct systems the heating must be in operation the summer and cooling in operation during parts of winter. Converting to VAV will allow turning this equipment off during these periods.

D. Pumps

1. If the heating equipment can be turned off in the summer and cooling equipment in the winter, so can their respective pumps, affording a excellent savings.

2. Smaller pumps and lower pumping capacities can be effected in VAV systems.

3. Variable volume pumps can be used.

E. Percentage Horse Power Savings with Different Methods of Fan Volume Control at an Average of 60 Percent of Peak Flow.

	Maximum Savings
Backward Inclined Fans with Discharge Dampers	13%
Airfoil or BI Fans with Inlet Vanes	36%
Forward Curve Fans with Discharge Dampers Located 3 fan diameters from fan	48%
Adjustable Speed Drive	50%
Forward Curve Fan with Inlet Vanes	57%
Adjustable Frequency AC Motor Control	78%

F. Ballpark Savings on VAV Energy Consumption

Total Building Energy Consumption, Per Cent Savings
Range ....20% to 50%
Average 35%

Total Building Energy Costs Savings, Per Sq Ft
Range ....$.30 to $.70
Average ....$.50

Fan Savings ....15% to 70%
Cooling Savings...30%
Savings on Pump Energy ....10% to 15%
Heating Savings...30%

G. Budget Costs For Conversion

* $400 to $1200 per box
*   35¢ to $1.00 per sq ft of building
* $200 to $600 per ton

PROCEDURE FOR EVALUATING A VAV CONVERSION

1. Calculate the new peak heating and cooling loads based on energy consumption measures planned or previously implemented and other possible system or building changes from original design. Distinguish true simultaneous peak building load from calculated building peak load, which assumes that the peak load occurs everywhere in a building at the same time. Note the diversity, which is the difference between the true simultaneous and calculated loads.

2. Determine lower Cfm required based on new heating and cooling loads and other changes. Note diversity in Cfm.

3. Before taking existing readings on fan:
    *   Make sure filters, coils, etc. are reasonably clean.
    *   Check system for imbalance. If excessive open all outlets 100 percent.
    *   Make sure automatic dampers are operating properly.
    *   Check for duct leakage. Correct if excessive.
    *   Make sure that the system is open and that no fire and manual dampers are shut.

4. Take readings of actual existing fan performance.
    *   Total Cfm Flow
    *   Suction and Discharge Fan Static Pressure
    *   Fan RPM
    *   Filter, Coil, Damper Pressure Drops
    *   Outside Air Flow and Operation
    *   Running Amps if In Question
    *   Record Motor Name Plate Data
    *   Check Starter and Overload Size

5. Determine operational hours of fan.

6. If possible put amp recorder on fan for 24 hours on a week day and weekend day.

7. Check the intake and discharge water temperatures, pressures and flows at coils.

If water flows and temperatures are excessively higher or lower than design, determine effect on air flow.

8. If working with DX coils, check if condenser system has unloaders, hot gas by-pass or multiple compressors to avoid freeze up at coils with lower air volumes.

9. Calculate new fan performance and drive sizes based on new peak load. See chapter 11, pages 11-3, 11-4.

10a. Select type of VAV system and VAV terminal units best suited to building and existing HVAC systems.

Types of VAV systems available are:

Cooling only, cooling with reheat coils Separate interior and perimeter systems Fan powered terminals with or without reheat coils, By-pass at terminal only or at both terminals and fan Dual duct VAV Induction VAV Riding fan curve.

See section in this chapter for details.

10b. The types of VAV terminals will correspond to the types of system listed above. Further considerations for terminal selection are dependent on types of controls. See sections on VAV terminals and controls in this chapter for details.

10c. Make sure the duct design and outlet air distribution is suitable at maximum and minimum flows. Change outlets as required.

11. Select type of fan volume control.

Fan volume control may be implemented with:
    Inlet or outlet dampers
    Variable speed motor controls
    Variable speed drives
    Variable pitch vane axial fans, Etc.
See section in this chapter for details.

12. Select type of control system best suited for new VAV system operation and compatibility with existing controls. See information

on VAV controls in another section in this chapter.

13. Calculate existing annual electrical consumption of fan motor in terms of kWh and costs.

$$kWh = \frac{1.73 \times Amps \times Volts \times Hours}{1000}$$

$$= \frac{1.73 \times 77 \times 460 \times 5260}{1000} = 322{,}070\,kWh$$

Old Costs kWh $\times$ Costs/kWh = 322,070 $\times$ \$.10 = \$32,545

14. **Calculate Fan Savings**
The following calculations use system S-1 from the sample audit on the Suburban Office Building in Chap. 4.

Savings by Reducing Maximum Cfm Load (See Chapter 11, p. 11-4, item 13.)

The maximum Cfm actually needed in the interior office areas served by system S-1 is a great deal less than being supplied. It can be reduced from about 1.2 Cfm per sq ft to .84 Cfm per sq ft. This reduces S-1 from 38, 000 Cfm to 26, 000 Cfm.

The fan savings due to this reduction are:

Existing   322,070 kWh
New   113,000 kWh
Savings   209,070 kWh

Fan Savings Due to VAV Operation During

The average reduction of air delivery in a during occupancy times is 40 percent of peak This reduces the maximum Cfm of 26, 000 for S-1 down to an average of about 15,600 Cfm which is a reduction 10, 400 Cfm.

The reduction in S-1 BHP, due to reducing the maximum Cfm, is from the existing 59 BHP to 19 BHP (see p. 11-3).

The converted VAV system will run at a peak of 19 BHP and at an average of 60 percent of this during the 2,800 hours of occupancy.

BHP Savings $= .6 \times 19 \text{ BHP} = 11.40 \text{ BHP}$
Converted to kWh $= .746 \times 11.40 = 8.5 \text{ kWh}$
Savings $= \text{kWh} \times \text{Hours} \times \text{Cost/kWh}$
                     $= 8.5 \times 2,800 \times .10$
                     $= \$2,380 \text{ per year}$

Savings by running VAV boxes at minimum during evenings and weekends during winter months. Units were shut off during non-occupancy during summer.

BHP Savings at Minimums $= .25 \times 19 \text{ BHP} = 4.75 \text{ BHP}$
Converted to kWh $= .746 \times \$1.4 = 8.5 \text{ kWh}$
Savings 4 winter months (115 hrs/wk × 17 wks = 1955 hrs)
kWh × hrs × Cost/kWh = 8.5 ×1955 ×.10 = $1,162

Figure 13-1
Cfm Savings with VAV System

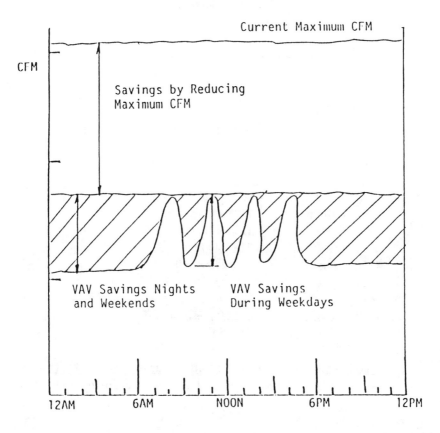

Figure 13-2
VAV Duty Cycle

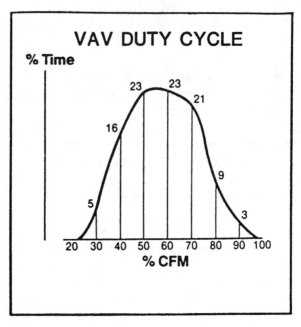

15. **Cooling Savings**
     <u>Cooling Savings by Reducing Maximum Cfm Loads</u>
     Savings in cooling is mostly a sensible energy reduction. This more
     conservative calculation of savings will be taken:

Cfm Reduction	= 38, 000-26, 000 = 12, 000 Cfm
Btuh Reduction	= 1. 08 × Cfm ×TD
	= 1. 08 × 12, 000 × (74-55)
	= 246, 240 Btuh
@ 60% avg load	= 246, 240 × . 60
	= 147, 750 Btuh (12-3 Tons)
	= 389 mill Btu
Dollar Savings	= @ $ 29 per million Btu for elect.
	= $11,281

<u>Cooling Savings by VAV Operation During Occupancy Times:</u>
Average reduction of 40 percent of peak load during occupancy,

Average Cfm Reduct.   = 26, 000 Cfm × . 4
                       = 10, 400 Cfm
                       = 1. 08 ×10, 400 ×19
                       = 213,400 Btu (17.8 Tons)
Occupancy Time   = 213,400 × 1400 hrs = 299 million Btu
                       = @ $29 per mill Btu electrical
                       = $8,671

16. **Recap of VAV Energy Savings**
     Fan Savings
         VAV Operation, Occupancy,   S-1 $2.380
                                     S-2 $2,380
     Cooling Savings
         VAV Operation, Occupancy,   S-1 $8,671
                                     S-2 $8,671
         Avoiding recooling   <u>$11,484</u>
                                                $33,586

17. **Estimate Retrofit Costs**
     Box Conversion to VAV
     Fan Volume Controls
     Controls
     Control Center
     Ductwork
     Wiring
     piping
     Total Costs $75,500 (See Chapter 9)

18. **Calculate Payback**

$$\frac{\text{Costs}}{\text{Savings}} = \frac{\$75,500}{\$33,586} = 2,25 \text{ years}$$

19. If the VAV conversion is feasible proceed with a detailed design,
     get quotations on equipment and installation and implement con-
     version.

20. Perform proper start up, check out performance, test and balance
     and monitor.

TYPES OF VAV SYSTEMS

Variable air volume systems are mostly cooling only systems most frequently used in interior areas that only require cooling. They are also employed because of their zone control capabilities. VAV systems can be low, medium or high pressure.

Understanding VAV Systems

Most HVAC systems in the past, with certain exceptions, have been constant air volume variable temperature type systems. A residential or small commercial system is typical, delivering for instance, a constant 1200 Cfm, while the burner or air-conditioner goes on and off changing the air temperatures to 58°F or 110°F to meet load conditions.

A VAV system is just the opposite. It delivers air at a constant 55°F while reducing or increasing the air quantities to satisfy changing space loads.

A true VAV system adjusts itself to changing solar, light, people, equipment, ventilation loads and diversities by increasing or decreasing the volume of air delivered to the spaces while still meeting desired comfort and health conditions.

As an average, VAV systems run at 70 percent of the peak load. Interior zones stay around 80 percent of maximum, plus or minus 10 percent. Perimeter zones vary more extensively in the cooling cycle because of shifting sun loads.

A VAV system becomes a constant volume system at maximum flow.

Types of VAV systems
1.   Cooling only system.
2.   Separate interior and perimeter system.
3.   Combined interior and perimeter system.
4.   Fan powered terminals with or without reheat coils.
5.   By-pass system at terminal units only or at terminals and fan.
6.   Dual duct.
7.   Induction.
8.   Riding the fan curve with VAV terminals.

COOLING ONLY VAV SYSTEMS

The second method is the true VAV or turn down VAV system. It actually throttles the air down at the terminal boxes rather than by-passing and cycling, and reduces the output at the fan, generally with inlet vanes on centrifugal fans or with variable pitch vane axial fans. Static pressure or constant volume monitoring stations in the main duct, monitor the changing air volume and pressure, and via controls, transmits a message to the fan directing it to increase or decrease its flow.

Figure 13-3
A true variable air volume system varies
the amount of air at both the terminals and fan

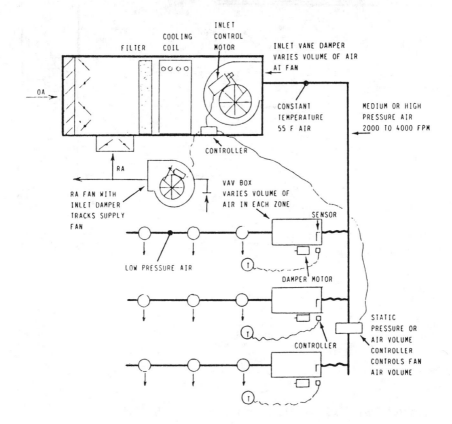

COMBINED INTERIOR AND PERIMETER SYSTEMS

Another approach is to combine the interior and perimeter terminals into one system. The interior cooling only system in this case is expanded to also handle the perimeter by adding perimeter reheat VAV boxes to it.

Figure 13-4
Perimeter VAV rehear boxes are combined
with cooling only interior VAV boxes on same system.

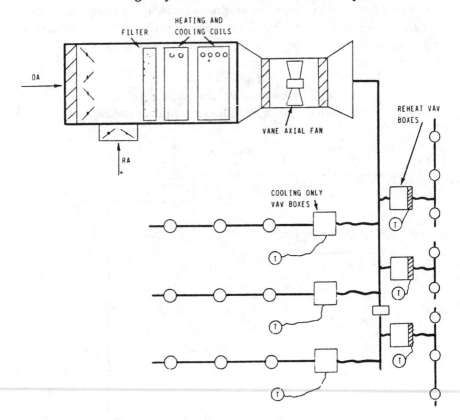

SEPARATE INTERIOR AND PERIMETER SYSTEMS

One common approach in buildings is to have separate interior and perimeter systems, such as a cooling only VAV system for the interior, and a heating only or heating and cooling system for the perimeter. The perimeter systems may be: low pressure constant air volume, hydronic baseboard, hydronic cabinets or induction.

Figure 13-5
Interior cooling only variable air volume system, combined with illustrations of four options for perimeter heating or heating and cooling.

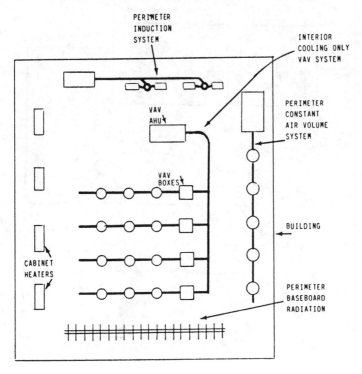

BY-PASS SYSTEMS

To vary the volume of air in the spaces, the by-pass system by-passes the supply outlets and dumps or cycles the excess air into the return air duct or plenum ceiling rather than actually throttling down the air. This is not considered a full true VAV system, but it works adequately for smaller simpler systems under 20,000 Cfm.

Figure 13-6
VAV boxes bypass supply outlets
and cycle air back to return system.

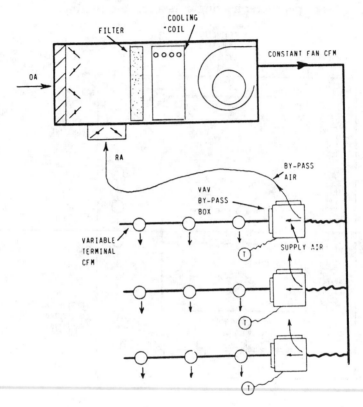

VAV TERMINALS

Types of VAV Terminals
1. Single duct cooling only.
2. Single duct with a reheat coil.
3. Fan powered.
4. Dual duct.
5. Induction.
6. By-pass.
7. System powered.

VAV boxes are classified by:
- • How they are powered: pneumatic, electric or system powered. Are they pressure independent or pressure dependent.
- • Are they normally open (N.O.) or normally closed (N.C.).
- • Are they controlled by direct acting or reverse acting thermostats.

Pneumatic and electric powered boxes employ separate pneumatic or electric motors to actuate the volume damper. System powered boxes use air from the high pressure duct system to power the flow dampers.

Figure 13-7
Typical pressure independent VAV box

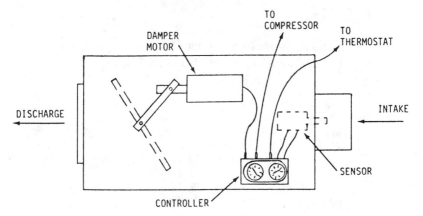

Figure 13-8
Converting CAV Reheat Box to VAV

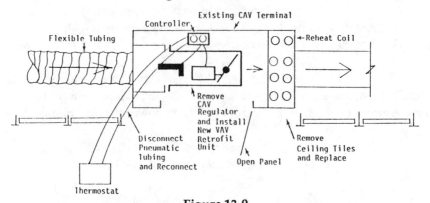

Figure 13-9
Typical Pressure Independent Volume Regulation
The variation in pressure between .3 inches and
4.0 inches is only about 5%

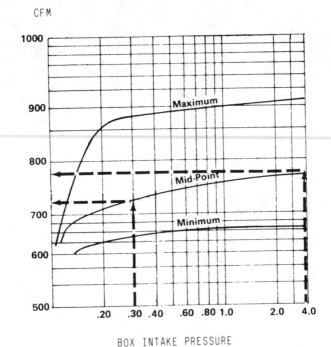

Figure 13-10
Typical Pressure Dependent Volume Regulation
At .03 inches 80 Cfm flows through box and at .3 inches 250 Cfm.

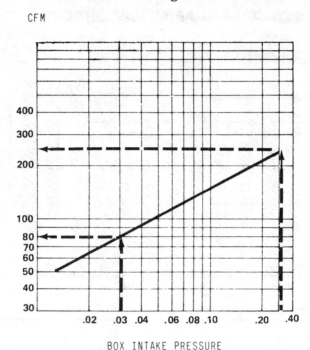

BOX INTAKE PRESSURE

METHODS OF VARYING FAN VOLUME IN VAV SYSTEMS

1. Inlet vane dampers on centrifugal fans.
2. Variable frequency motor speed controllers.
3. Variable pitch motor sheaves.
4. Variable pitch van axial fans.
5. Internal centrifugal wheel shrouds.
6. DC motors.
7. Eddy current couplings.
8. Fan by-pass.
9. Ride fan curve of forward curve wheel.

Figure 13-11
Energy Used After VAV Retrofit

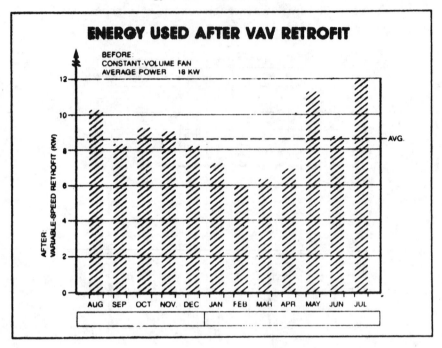

A 20-hp constant-volume system operating at 18 kW was retrofitted to VAV with a variable-speed drive, which resulted in the average monthly power reduction shown. Based on data logged, the annual reduction was 36,000, kWh, a 58% reduction

Figure 13-12
Percent Horse Power Saved
Percent drive power consumption

PERCENT DRIVE POWER CONSUMPTION

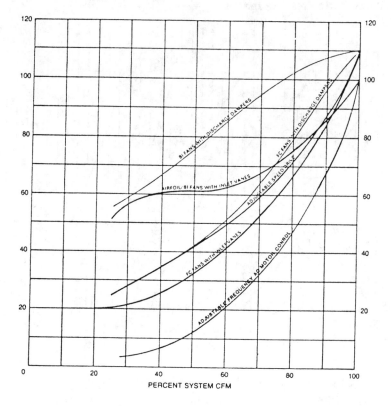

PERCENT SYSTEM CFM

Figure 13-13
Inlet Vane and Motor Speed Controller BHP Comparison
Total Static Pressure in. WG

Total Static Pressure In. WG

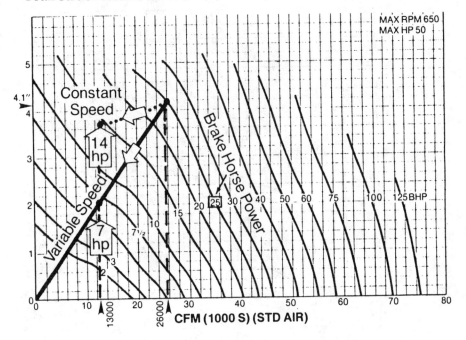

CFM (1000 S) (STD AIR)

At 26,000 Cfm on a centrifugal fan with a BI wheel there is a 25 HP draw. Reducing the fan flow in half with inlet vane damper blade to 13,000 Cfm at a constant speed, only reduces the HP to 14. With variable speed motor controls the horsepower reduces to 7.

Figure 13-14
Typical Inlet Vane Performance
Cfm in thousands

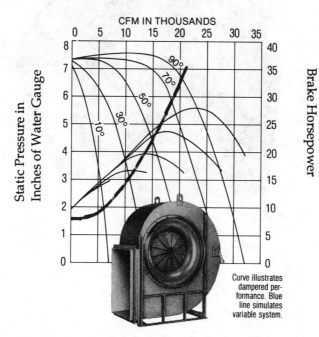

Curve illustrates
dampered per-
formance. Blue
line simulates
variable system.

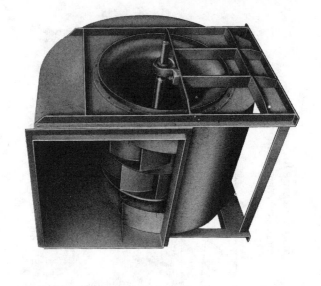

Figure 13-15
Typical Internal Wheel Shroud
Percent SP, BHP, and ME

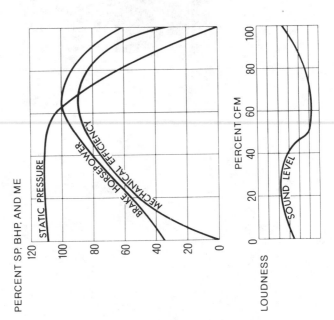

Figure 13-16
Riding The Fan Curve

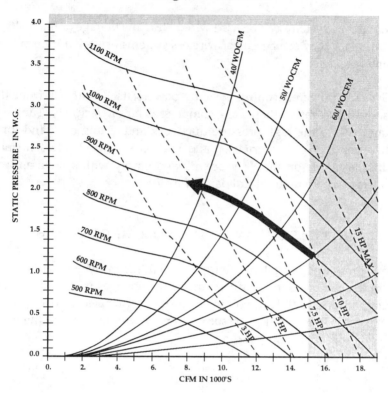

BASIC TYPES OF CONTROL SYSTEMS

There are various types of control systems used in HVAC systems.

1. Smaller buildings and systems generally use conventional indirect electric controls with low voltage wiring between VAV boxes and thermostats, etc.

2. Larger buildings and systems in the past years generally have used indirect pneumatic controls with a compressor.

3. Direct digital controls (DDC) electronic systems are being used more widely today. They can control dampers, valves, start-stop functions, sensing temperatures, flows, pressures etc. all directly. They also can

modulate dampers, valves etc. smoothly with instantaneous proportionate control.

4. Electronic systems can be hard wired with low voltage or line voltage wiring, or they can be carrier wireless systems using FM transmitters and receivers.

5. Systems powered control systems powers itself with air from the supply duct rather than using compressed air or electricity. Tubing is tapped into the high pressure ductwork and strung back and forth between the space thermostat and VAV box. Two bladders, a small and large one on the VAV box fill and empty with system air and operate the volume control damper inside it.

TWO BASIC TYPES OF VAV SYSTEM VOLUME CONTROLS

1. <u>Static Pressure Controllers</u>—Roughly two thirds the way down the duct system a static pressure sensor monitors the changes in pressure in the system and directs the fan volume controller to increase or decrease air volume accordingly. Static pressure sensors are set to measure the minimum pressure in the longest run.

Figure 13-17
DWDI Air Foil Blower

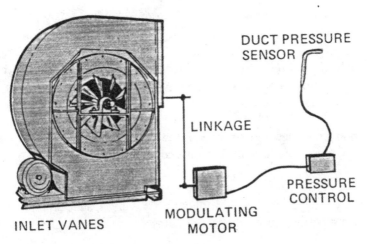

2.   **Volume controls** Air measuring stations with pitot type type sensors can be incorporated at various locations in the air distribution system such as at the discharges of the supply and return fans to measure and modulate supply and return fan volumes.

Figure 13-18
Volume Controls

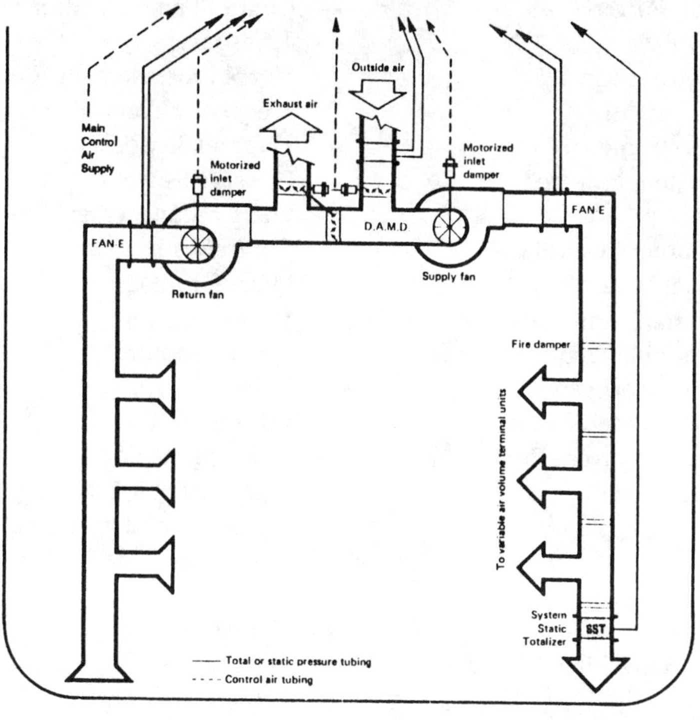

VAV CONVERSIONS
Check Off List

VAV Boxes
— Cooling Only
— Cooling with Reheat Coils
— Induction
— Dual Duct
— Fan Powered
— By Pass
— Retrofit Kit
— Pneumatic
— Electronic
— System Powered

Fan Volume Control
— Inlet Vane Dampers
— Variable Pitch Sheaves
— Variable Frequency AC
  Motor Speed Controllers
— Fan Wheel Shrouds
— Eddy Current Couplings
— DC Motors

Controls
— System Static Pressure Sensor
— System Static Pressure
  Transmitter
— Air Flow Measuring Stations
— Control Panels, Volumetric
— Transducers
— Signal Processors for Variable
  Frequency Controllers
— Thermal Sensors
— Pressure Sensors
— Energy Management System
  Control Unit
— Pilots for Inlet
  Vane dampers
— Control Motors
— Tubing
— Wiring
— Thermostats
— Remove Tubing, Wiring
— Reinstall Tubing, Wiring

Fans   Testing
— Centrifugal
— Variable Pitch Vane Axial

Ductwork —
— Low Pressure Galvanized
— High Pressure Galvanized
— Flexible Tubing
— Sheet Metal Accessories
— Ductwork Insulation
— Remove Ductwork
— Reinstall Ductwork
— Flexible Tubing
— Insulation

— Old Fan Readings
— Old Box and Outlet Readings
— New Fan Readings
New Balancing Boxes
  & Outlets
— kW Metering

Piping
--- Remove Piping
— Reinstall Piping
— BI Piping
— Copper Tubing
— Control Valves
— Valves
— Insulation

Section 8

Indoor Air Quality

Chapter 14 -

Indoor Air Quality Retrofits

Chapter 14

Indoor Air Quality Retrofits

PROBLEMS

Exposure to indoor air pollution is one of the major environmental health hazards in the United States today. IAQ problems lead to worker complaints, health problems, respiratory problems, reduced productivity and lawsuits. IAQ related health problems are estimated at $15 billion a year and almost 2 billion lost work hours. About 30 percent of the people are complaining. Allergies caused by airborne biological particulates are estimated to effect 25 million people in the United States.

Indoor pollutant levels are often higher than those outdoors. Generally outdoor air is 20 times cleaner than indoor air and the average American spends 90 percent of his or her time indoors. Organic pollutants may be two to five times higher inside homes and individual chemical pollutants up to 20 percent higher than outside.

Up to 20 percent of new commercial buildings may be affected by unhealthy concentrations of organic compounds as much as 100 times higher than found outdoors.

There is an imminent need for cleaner indoor air, to know and recognize causes, to eliminate the IAQ problems and pollutants, and to be able to diagnose, test and rectify effectively. Knowledge and skills with instruments, IAQ auditing procedures, analyzing skills, methods of correcting HVAC system deficiencies etc. are required by those involved with IAQ.

HOMES MORE POLLUTED THAN SOME FACTORIES

People breathe in three times as many carcinogens inside their own homes than on the street, even if they live in an industrial area. The effect is even worse for the children of smokers, who run twice the risk of developing leukemia if one of their parents smokes and four times the risk if both are smokers, said Lance Wallace of the Harvard School of Public Health.

In 1981, researchers equipped more than 350 residents in New Jersey with monitors to test the air they were breathing for a number of different chemicals. There was no difference between the people living close to petro-chemical plants than farther away. Wallace said the levels of cancer causing agents was two to five times higher indoors than outdoors, and in some cases, there was 100 times more of a particular compound inside the house than in the backyard.

Common household items such as cleaners, pesticides and room deodorizers were the source of much of the pollution, Wallace said.

OSHA levels are concerned more with dangers from immediate rather than prolonged exposure. Chronic exposure is more a factor in cancer development, Wallace said, the homes of smokers being the best example.

Cigarette smoke contains benzene, a known cause of leukemia, Wallace said. Smokers homes were found to contain 30 to 50 percent more benzene than other homes.

What is Good IAQ?

Good IAQ must be conducive to the health and well being of the occupants. Pollutants must be either removed or kept down to acceptable levels.

Concentrations of oxygen and carbon dioxide must be within acceptable ranges to allow normal functioning of the respiratory system. Concentrations of gases and vapor, biological and non-biological particulates, and radio nuclides must be below levels that are harmful or can be detected as objectionable by the occupants. Because some contaminants-such as carbon monoxide from combustion, asbestos from insulation, radon from soil, and polonium from tobacco smoke, are odorless but can have life-threatening effects, sensory responses may not adequately protect individuals from exposure to air pollutants.

Retrofit Opportunities

With three out of four sick buildings directly attribute to inadequate ventilation, and with building owners more knowledgeable and receptive to IAQ issues, a tremendous retrofit business opportunity is developing. Advantages can be taken of low cost, easy to use diagnostic instruments and retrofit options.

CAUSES OF POOR IAQ

General Causes

Investigations by the National Institute for Occupational Safety and Health and other government and private sector agencies over the past decade help put the problem in perspective. A summary of specific causes of SBS provided by NIOSH and its Canadian counterpart, Health and Welfare Canada, is presented in Table 1. The findings of both government agencies are almost identical.

In 52% of their investigations, inadequate ventilation such as insufficient fresh air intake and low ventilation effectiveness was identified as the principal causal factor.

Another 12% to 16% of IAQ problems were related to indoor generated contaminants including tobacco smoke.

Nine to 10% were linked to infiltration of outdoor contaminants such as improper intake air locations, another ventilation-related factor.

Other identified factors include contamination from building fabric and materials (2% to 4%) and microbial problems (0.4% to 5%). The cause of IAQ problems could not be determined in 12% to 24% of the investigations.

A closer look at the investigative data reveals that indoor contaminants account for less than one in four cases where problems exist. Instead, three out of four sick buildings are directly attributable to inadequate ventilation.

Table 1
Causes of IAQ Problems

Problem Type	NIOSH Survey	HWC Survey
Inadequate Ventilation	52%	52%
Indoor Contaminants	16%	12%
Outdoor Contaminants	10%	9%
Building Fabric	4%	2%
Biological Contaminations	5%	0.4%
No Problem Found	12%	24%

Measuring CO_2 levels are a good indicator of the amount of fresh air being brought indoors and can help determine ventilation problems and needs.

SPECIFIC CAUSES OF INDOOR POLLUTION

Inadequate Ventilation
- Inadequate Outside Air
- Improperly Controlled Outside Air
- Inadequate Exhausts Toilet, Kitchen, Return Air, Fume etc.
- Tight Building
- Short Circuiting of Air In Spaces
- Improperly Pressurized Building
- Inadequate Filtering of Air

Indoor Generated Contaminants
- Tobacco Smoke
- Gas Leaks
- Freon Leaks
- Aerosols, Cleaners Hair Sprays Cleaning sprays Disinfectants
- Pesticides
- Fumes, Chemical
- Nitrogen Dioxide
- Products of Combustion, Carbon Monoxide Holes in Heat Exchanger and Flues Clogged Flue or Chimney Engine Exhausts, cars, trucks
- Copy Machines
- Lasers
- Habachis and Charcoal Broilers

Building Materials, Fabrics, Furnishings
- Carpets
- Sheets and Blankets
- Carpet Adhesives
- Furniture
- Fabrics in Furniture and Drapes, etc.
- Wood, Plywood
- Insulation
- Paneling, Particle Board
- Plastics, Laminates
- Asbestos

Indoor Biological Contamination
- Carbon Dioxide CO_2 from People
- Pollen
- Mold
- Fungi

- Mildew Spores
- Standing Water
- Cooling Coils, Drain Pans
- Dust Mites
- House Dust
- Animal Dander
- Bacteria and Viruses
- Humidifiers (not evaporative type)

Outdoor Contaminants
- Due to Infiltration
- Radon
- Soil Gas
- Methane
- Pesticides
- Auto Pollution
- Exhaust Stacks
- Due to Outside Air Intake Drawing In Contaminated Industrial Process

Industrial Indoor Contaminants
- Paints
- Chemicals
- Printers
- Particulates, small solid or liquid particles such as dusts, powders, liquid droplets and mists. Examples, fly ash and asbestos dust.
- Gas Pollutants, fluids without form that occupy space rather uniformly such as carbon monoxide or chloroform.
- Fumes, irritating smoke, vapor or gas.
- Pollutants maybe toxic, noxious, corrosive erosive, inflammable, explosive or radio active.

TESTING, SETTING AND CONTROLLING OUTSIDE AIR

Correctly and consistently controlled minimum outside air volumes is an absolute requirement for good IAQ and can be achieved as follows:

Outside Air Needed for IAQ
Provide sufficient outside air to meet codes, healthy ventilation, building pressure and direct exhaust makeup air requirements. Addi-

tional outdoor ventilation air has been shown to be the single most effective method of correcting and preventing problems and minimizing complaints related to poor indoor air quality. Even if a specific contaminant is identified (such as formaldehyde) dilution may be the most practical way of reducing exposures.

Residences:	0.35 air changes per hour
	This is comparable to 15 Cfm
	per person.
Classrooms:	15 Cfm per person
Offices:	20 Cfm per person
Public restrooms:	50 Cfm per person
Where smoking is permitted:	60 Cfm per person

This equates to generally 15 to 20 percent of total Cfm for a 10,000 Cfm system in a 10,000 sq ft office building which may need a minimum of 1500 to 2500 Cfm. This equates to

.15 or .20 Cfm per sq ft of building.

The amount of fresh air naturally infiltrating into a tightly insulated home without ventilation may be as low as a hundredth (.01) of an air change per hour.

Check Out Actual Outside Air Conditions

1. Measure actual amounts of outside air being taken in under different outside temperature conditions and building load conditions.

2. Check if minimum air volumes are correct and being held. During occupancy periods should not close beyond the minimum position and fans and air, handling units should run continuously.

3. During occupancy periods should not close beyond the minimum position and fans and air handling units should run continuously. Check if correct amounts of OA is being taken in at maximum OA settings.

4. Check if sufficient amounts of cool outside air is being taken in when required for cooling.

5. Check winter, spring, summer and fall conditions.

6. Adjust damper linkages as required and reset controls or change controls as required.

7. Check that all supply outlets in spaces are opened to their correct balance positions.

Constant Mixed Air Controller May Not Work

The typical older control system with a constant mixed air temperature controller with a sensor in the mixed air plenum does not control the volume well, because it generally exceeds minimum OA requirements, even though it can also go under minimum air requirements. The percentage of outside air coming in under the minimum mode may vary from 10 percent to 100 percent depending on the OA temperature.

For example, the following percentages of OA are required to maintain a fixed 55°F mixed air temperature with a constant 70°F return temperature with a constant mixed air temperature controller.

Outside Air Temperature, F	Percent OA Required to Maintain 55°F
–10°F	19%
0	21
10	25
20	30
30	37
40	50
50	75
55	100

The average percentage of outside air intake during the winter season for various cities is:

Chicago,	42 percent
Boston,	47 percent
Denver,	44 percent
Atlanta,	54 percent

The solution to this problem of the constant mixed air controller not controlling the volume of outside air correctly, is to move the constant mixed air temperature controller to the leaving side of the heating coil and use it as a low limit temperature controller instead.'

Must have automatic changeover control in economizer cycle to switch back and forth between minimum outside air damper settings and the settings for using outside air as cooling.

What to Watch Out For
1. Don't reduce outside air volume below direct exhaust air quantities thereby putting the building under a negative pressure and forcing air infiltration.

2. Maintain some building pressurization. This means bring 1 to 5 percent more outside air than is exhausted.

Common Outside Air Plenums
If working with a common outside air plenum, put separate OA dampers on intake of each HVAC unit. The outside air plenum itself can be maintained as is.

METHODS OF OUTSIDE AIR CONTROL

1. Low Limit Control—on leaving side of heating coils is an accurate method. High Limit Constant Mixed Air Temp—based on mixed air temperature and mixed air controller is not an accurate method.

2. Mixed Air Reset—A constant minimum outside air volume is maintained when outside air is not used for cooling.

3. Outside Air Volume Measuring Station—Air measuring station constantly measures outside air volume and automatically controls minimum flow and economizer flow very accurately.

4. Manual Minimum Outside Air Control—Outside air damper opens up automatically to fixed minimum position when system is in operation and closes when system is turned off.

5. Separate Outside Air Fans—used exclusively for bringing in outside air.

CARBON DIOXIDE LEVELS

Measuring CO_2 levels are a good indicator of the amount of fresh air being brought indoors and can help determine ventilation problems and needs.

ASHRAE recommends indoor levels of CO_2 kept below 1000 ppm.

CO_2 levels increase with occupancy because people exhale CO_2.

Outdoor levels of CO_2 typically range from 320 to 350 ppm.

Infrared technology now makes the diagnosis of ventilation adequacy easy. Low cost ventilation efficiency measurement systems can quickly generate a ventilation record tracking CO_2 levels over time as a ventilation index. ASHRAE

Today's technology provides a number of possible low cost product solutions to ventilation adequacy, including ventostats (CO_2-based ventilation controllers), desiccant wheels, energy recovery equipment, etc. Demand Control Ventilation is one of a number of control strategies that can be used.

Figure 14-1
Ventograph
Doctor's Waiting Room — Phoenix, Arizona

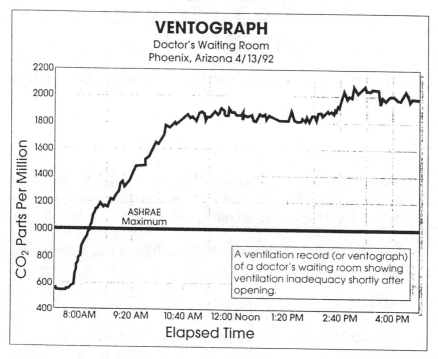

The ventograph above tracks CO_2 levels in a doctors waiting room.

Figure 14-2
Ventograph
Airplane Flight — Tampa to Cincinnati

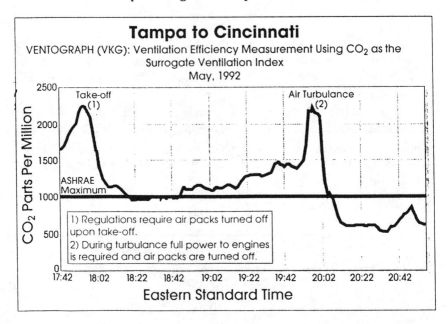

The ventograph above records air quality levels during a trip on a commercial airliner.

INDOOR AIR QUALITY INSTRUMENTS AND TESTING

1. A soup to nuts indoor air quality inspection of a typical 100,000 square foot office building could cost $5,000 to $10,000 depending on the complexity of the testing and conditions.

2. Building owners and managers should audit their IAQ at least once a year.

3. IAQ audit

    Talk with building owners, maintenance staff and occupants

    Study HVAC systems to find common problems

Check for adequate outside air. Check that outside air dampers are open. They may have been closed deliberately or by a faulty temperature control system. Check that they maintain the minimum amount of outside air flowing during occupancy and that they operate properly through.

Most problems can be found in a day long inspection.

4.   Make a log of employee complaints.

5.   Check ventilation systems plus temperature and humidity factors.

6.   Seek and evaluate sources of contaminants.

INDOOR AIR QUALITY INSTRUMENTS AND TESTING

There are a multitude of excellent testing instruments on the market, many for very reasonable prices. Industrial Hygienists, HVAC Balancing Technicians and HVAC Engineers may already own or be familiar with many of them.

- Carbon Monoxide, CO Analyzers
- Carbon Dioxide, CO_2, Analyzers
- Radon Detectors, RN
- Microbiological Samplers
- Tobacco Smoke Testing
- Volatile Organic Compound, VOC Samplers
- Semi-Volatile Organic Compounds
- Bioaerosol Testing
- Air Flow Measuring Instruments such as Air Flow Hoods, Pitot Tubes and Anemometers, Hot Wire Anemometers etc.
- Mold, Yeast, Fungus Tests
- IAQ Testing Kits
- Tests for Lead and Lead Dusts
- Tests for Nicotine
- Ozone Test
- Tracer Gas Instrumentation
- Natural Gas Test
- Temperature recorders
- Hygrometer for Humidity Measurements

- Refrigerant Leak Detectors
- Testing for Bioaerosols
- Instruments for Formaldehyde Testing

SOLUTIONS AND CONTROL OF INDOOR AIR POLLUTION

General Control Methods
The three general ways to combat indoor air pollution are:

- Source control: May involve isolation, product substitutions and local exhausts.

- Dilution: Involves infiltration, natural ventilation and mechanical ventilation.

- Removal: Includes fan-filter modules, clean benches and central forced air systems with recirculated air.

1. Clean Ducts
2. Clean and replace filters periodically.
3. Clean clogged grilles, filters and dampers.
4. Clean drain pans on discharge of cooling coils.
5. Balance air distribution systems if out of balance.
6. Make sure drain pans drain all water out properly.
7. Make sure humidifiers are controlled properly. Are they breeding bacteria, wrong type, etc.
8. Make sure the proper amount of outside air is coming in at all times for good ventilation and that it is controlled correctly under different circumstances.
9. If building is under negative pressure analyze building air balance and reverse to Pressurizing building.
10. Clean carpets infested with microorganisms, dust mites. Spray with a disinfectant.
11. Exhaust, dilute and contain cigarette smoke.
12. Check and correct inadequate fume exhausts from burners. See that there is no backup of the products of combustion.
13. Check for holes in heat exchanges and flues. Check for clogged chimneys or those not high enough.
14. Where room layouts have been changed, partitions moved, etc., check that air distribution has been updated accordingly.

15. Check for natural gas leaks.
16. Are toilet exhaust systems doing their job.
17. Check fume exhaust stations, hoods, fans that they are in proper operation, etc.
18. Is there adequate makeup air coming in the building.
19. Test for radon leakage from the earth.
20. Make sure kitchen exhaust systems, hoods, fan, filters, work effectively, are clean and operate properly.
21. Increase Relative Humidity to reduce respiratory illnesses in the winter.
22. Dust Mites
23. Keep HVAC equipment and controls maintained.
24. Baking New Buildings Temperature is raised to 90°F or more while ventilation equipment is run full speed to exhaust compounds given off by new materials in the building and its furnishings.
25. CO_2 based ventilation controllers

Humidity

The ideal humidity guideline should specify a relative humidity range that minimizes deleterious effects on human health and comfort as well as reduces, as much as possible, the speed of chemical reactions of the growth of biological contaminants (which will impact human health and comfort).

Like most gaseous and particulate contaminants, relative humidity is primarily affected by indoor and outdoor sources and sinks. However, unlike other contaminants, relative humidity is also a function of air temperature.

In addition to the effect of temperature, selecting the most desirable range of humidity is complicated by the conflicting effects of an increase or decrease in humidity levels. For example, while increasing humidity may reduce the incidence of common respiratory infections and provide relief for asthmatics, an increase in humidity may also increase the prevalence of microorganisms that cause allergies. Criteria for indoor exposure must balance both effects.

The following figure graphically summarizes the apparent association between relative humidity ranges and factors that affect health of occupants at normal room temperature. The figure is constructed as a bar graph relating relative humidity from 0% to 100% (shown along the horizontal axis) to:

Figure 14-3
Optimum Relative Humidity Ranges for Health

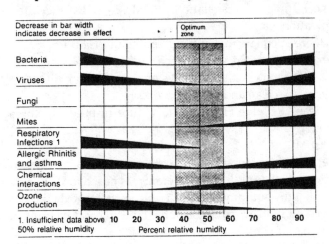

- Biological organisms (bacteria, viruses, fungi and mites);
- Pathogens causing respiratory problems (respiratory infections, asthma and allergies); and
- Chemical interactions and ozone production.
  The decreasing width of the bars represents decreasing effects.

The bacterial population increases below 30% and above 60% relative humidity. The viral population increases at relative humidity below 50% and above 70%.

Fungi do not cause a problem at low humidity. However, their growth becomes apparent at 60%, increases between 80% and 90%, and shows a dramatic rise above 90%.

Mites require humidity for survival. Growth in the mite population responds directly to humidity levels in excess of 50%. The incidence of allergic rhinitis because of exposure to allergens increases at relative humidities above 60% and the severity of asthmatic reactions increases at relative humidities below 40%.

Most chemical interactions increase as the relative humidity rises above 30% although ozone production is inversely proportional to the relative humidity.

The evidence suggests that the optimal conditions to enhance human health by minimizing the growth of biological organisms and the speed of chemical interactions occur in the narrow range between 40% and 60% relative humidity at normal room temperature. That narrow range is represented by the optimum zone in the shaded region of the graph.

Although keeping indoor humidity levels within this region will minimize health problems, there is probably no level of humidity at which some biological or chemical factor that affects health negatively does not flourish.

PARTICULATE CONTAMINANTS

Particulates

Indoor air particulates may come from outdoor sources or indoor sources. Particulates are usually categorized according to size:

1. Respirable, less than 5 to 10 micrometers diameter which can lodge in the lungs and cause definite health problems.
2. Nonrespirable, greater than 5 to 10 micrometers diameter.

Typically reported range for total particulates of all sizes is 300 to 1000 micrograms per cubic meter averaged over 24 hours, with maximum readings of 600 micrograms per cubic meter. The indoor/outdoor ratio typically varies from 0.3 to 0.4. The following standards have been established:

Tobacco Smoke

Hypersensitivity to tobacco smoke is fairly common, often resulting in irritation of the eyes and respiratory tract.

Tobacco smoke consists of solid particles, liquid droplets, and gases, and constitutes more than 2,000 specific materials. Like many pollutants, it can be absorbed by the body unsuspectingly. A study in Britain revealed traces of tobacco substances in the urine of a test group of nonsmokers in 85 percent of the cases, despite the fact that half of the group's members were unaware that they had been exposed to a low-level dose of tobacco smoke.

Fungal Spores

Fungal spores are a broad class of biological organisms that can function as potential allergenic agents. Those most commonly found indoors are associated with mildew and decay and can be found in air conditioning systems. Fungal spores can also originate outdoors, where their numbers are subject to seasonal variations. Spore concentrations outdoors rarely exceed 1 spore/cc, and normal dwellings are likely to have lower concentrations.

Fibers

Fibers can include several types of mineral or organic fibrous material. The most important of these from a health standpoint is asbestos, and only this type of fiber is considered here. Asbestos can occur in many forms, including amosite, chrysotile, and crocidolite. It may be found indoors through its use as a construction material (insulation), although this use has been severely curtailed in recent years.

Asbestos fibers, when lodged in the lung, can cause asbestosis (a disease of the lungs) and mesothelioma (a cancer that attacks the lining of the chest cavity or abdomen).

OSHA is using <u>0.1 fibers</u> longer than five micrometers per cubic centimeter as the level above which abatement action must be taken.

The following standards have been established:

<u>OSHA</u> 2.0 fibers/cc (8-hour TWA for industrial exposure)
<u>ACGIH</u> 0.5 to 2.0 fibers/cc (8-hour TWA for industrial exposure)
<u>ASHRAE</u> lowest feasible level

Pathogens and Allergens

Pathogens are particulates that cause disease. Allergens, similarly, induce allergies. Buildings harbor both in the form of bacteria, viruses, mold spores, pollens, insect parts, and people and pet materials. Often, these particles adhere to other particles, primarily dust, and collect unseen in carpets, fabrics, duct systems, humidifiers, fan coil units, cooling towers, etc.

Exposure to such contaminants can induce a variety of allergic reactions and illnesses: cold, flu and several forms of pneumonia, including Legionnaires' disease. Tuberculosis, measles, small pox, staphylococcus infections, and influenza are known to be transmitted by air.

Biologic Aerosols

Increasing attention to biologic aerosol components of indoor air has resulted from investigations that have shown that airborne concentrations of viable organisms frequently correlate with physiologic responses and complaints. Symptoms including pulmonary manifestations, muscle aches, chills, fever, headache and fatigue have been attributed to biologic agents.

Disease has been attributed to thermophilic actinomycetes, non-pathogenic amoebae, fungi, and Flavobacterium spp. or their endotoxins.

Polyaromatic Hydrocarbons

They are produced indoors by incomplete inorganic combustion including tobacco smoking, wood-burning, and cooking. Polyaromatic hydrocarbons occur in vapor and condensed forms. Usually, only the particulate (condensed) form is measured.

Complex Mixtures

Cooking by-products, combustion products, tobacco smoke, exhaled human breath, and other indoor air constituents are composed of complex mixtures often including varieties of organic and inorganic vapors as well as particulates. These mixtures change over time and through space after generation as they mix with room air, combine chemically, and undergo physical changes.

Gaseous Contaminants

Carbon Dioxide, CO_2
Carbon Monoxide
Freons
Nitrogen Dioxide
Total Hydrocarbons
Sulfur Dioxide
Ozone
Formaldehyde

Organic Compounds

The following are ACGIH limits for some organic contaminants commonly found in office buildings:

Compound	ppm (TWA)
Styrene	50
n-heptane	400
Toluene	100
Octane	300
m, p-xylenes	100
o-xylenes	100
ethyl benzene	100
Nonane	200
Cumene	50

Formulas

TLV: stands for threshold limit value. It's the time weighted concentration that normally healthy adults can withstand for eight hours a day (40 hours a week) without adverse effects. It is usually stated in parts per million (ppm).

TLV-STEL: the maximum concentrations to which workers can be exposed for up to 15 minutes.

TLVC: is the level that cannot be exceeded even instantaneously.

Effective temperatures are sometimes used to express the effect that dry-bulb temperature, wet-bulb temperature, and air velocity have on comfort. It is a single-number index.

WBGT: is the wet-bulb globe temperature, which includes the radiant effect.

Dilution Air

$$\text{Cfm} = \frac{403 \times \text{SG} \times 10^6}{\text{MW} \times \text{TLV}} \times \text{pints/min} \times K$$

Where:

SG   = specific gravity

MW   = molecular weight

K   = a safety factor that varies from 3 to 10 depending on the toxicity of the material and the effectiveness of the ventilation. This is difficult to estimate. We will use two values, K1 (the value for toxicity) and K2 (the value for ventilation), which together add up to K.

Fire and Explosion

$$\text{Cfm} = \frac{403 \times (\text{S.G}) \times (100)\,(C)\,(\text{pints/min})}{\text{MW}\,(\text{LEL})\,(B)}$$

Where:

LEL   = Lower explosive limit (given as a percentage) which is expressed in parts per hundred.

MW   = molecular weight of liquid.

SG   = specific gravity of liquid.

C   = safety factor, which depends on the percentage of the LEL necessary for safe operation because concentrations are not uniform. For continuous ovens, C= 4, and for batch ovens, C=12. B = a constant that takes into consideration the fact that LEL decreases at elevated temperatures. B =1 at temperatures less than 250°F, and 0.7 at temperatures equal to or more than 350°F.

Part II

How to Estimate and Retrofit
HVAC System Components

Section 9

Heating and Cooling Equipment

Chapter 15

Heating Equipment

There are a host of valuable methods for saving heating energy. Improving combustion efficiency can easily save 10 percent, for example. Staging heating, resetting boiler discharge temperatures, installing automatic vent or draft dampers, switching to multiple or new boilers and putting more insulation on equipment are more of the methods available. Also using more efficient fuels and burners, and controlling feed water, steam systems, traps etc. are ways to save heating energy.

This chapter reviews these methods again and provides price and labor tables for replacing heating equipment.

HEATING ENERGY IMPROVEMENTS

Combustion Efficiency
- Check combustion efficiency and adjust. Read stack temperature and CO_2 or O_2 and tune up. Reduce stack temperatures and minimize excess air.

- Install automatic oxygen trim controller to maintain maximum combustion efficiency at all times.

- Clean soot, scale etc. from tubes and firewalls.

- Schedule boiler blow down on as required basis rather than on a regular schedule.

- Check boiler areas for building negative pressure. Correct to provide proper amount of combustion air.

255

Staging Heating
- First burner on at 50°F outside.

- Second burner on at 30°F outside.

- Third burner on at 10°F outside.

Reset Temperatures
- Reset boiler hot water discharge temperature proportional to outside air temperature. If the outdoor temperature is 50°F or 60°F lower hot water temperature to 120°F. If outdoor temperatures are 0°F or 20°F raise temperature to 160°F.

- Install automatic temperature reset controls.

Vents, Drafts
- Install automatic draft or vent damper to reduce stack losses when boiler is not firing.

Multiple and New Boilers
- Install either multiple boilers in place of one operating excessively at part load or a single smaller boiler.

- Operate one boiler to its maximum load before bringing other boilers on line.

Insulation
- Insulate boilers, HVAC units, reduce radiation losses with burners etc. which are in unheated spaces, outside on roof or in air conditioned spaces.

Heat Recovery Equipment
- If stack temperatures exceed 300°F install a heat recovery unit in the stack and use reclaimed heat to heat up make up air, combustion air, domestic hot water or for space heating.

- Install turbulators or baffles to improve heat transfer if stack temperatures are too high (over 450°F) and combustion efficiency is at a maximum.

Burners and Type Fuel
- Convert to more efficient fuel

- Install new, more energy efficient burners

- Install electronic ignitions in place of standing pilot lights.

Feedwater
- Install an economizer for preheating boiler feed water.

- Maintain proper level of additives and chemicals.

Steam
- Clean steam traps

- Seal leaky steam traps

- Monitor steam trap leakage

- Keep steam pressure as low as feasibly possible for coils or radiation equipment.

- Vary steam pressure proportional with demand.

- Install steam pressure regulators to reduce steam usage. Use by passes only for repairs and emergencies.

- Install steam flow meter on larger systems.

- Install heat recovery units in condensate pump systems to reclaim heat and reduce temperature of the condensate.

- Use blowdown heat recovery

Oil
- Install automatic viscosity controls on fuel oil systems for maximum atomization and the ability to use other grades of fuel etc.

- Reduce firing rates of oil or gas burners if boiler is oversized to avoid short cycling.

STEAM TRAPS

The proper operation of steam traps can greatly influence the overall efficiency and energy consumption of a boiler system. Steam traps remove condensate, air and carbon dioxide from steam systems. Over a period of time internal parts wear and the trap fails to open and close properly. Several different tests may be applied to check their proper functioning.

1. Listen to the trap to check if it is opening and closing on a definite cyclical basis. This test may better be performed with an audio stethoscope specially designed for this purpose but can also be done, with a somewhat reduced sensitivity, with the use of a screwdriver pressed to the ear.

2. Feel the pipe on the downstream side of the trap.

    •   If it is as hot as the upstream side, the trap may be passing steam. This condition can be caused by dirt caught in the trap seat, valve or stem, by excessive steam pressure or worn trap parts (especially valve and seats).

    •   If it is moderately hot, as hot as a water pipe, for example, it probably is passing condensate, which is proper.

    •   If the downstream side is cold, the trap is not working at all.

3. Measure temperature drop across the trap with a surface pyrometer.

    •   A normally operating trap should show a temperature difference of 10 to 20 degrees F across the trap.

    •   Lack of drop usually indicates steam blowthrough.

    •   Excessive drop indicates that the trap is not passing condensate.

4. Check back pressure on downstream side. Measure temperature of return lines.

All traps should be inspected on a regular basis and faulty traps should be adjusted, repaired or replaced as necessary.

Calculating the Cost of a Leaky Steam Trap

1. One trap in a 30 psig system with a 0.125 inch orifice loses 15 pounds

Figure 15-1
Steam Trap Loss Nomograph

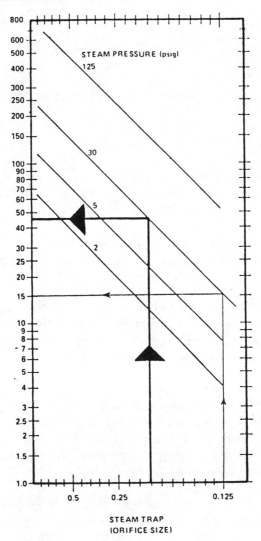

STEAM LOSS
(POUNDS PER HOUR)

STEAM TRAP
(ORIFICE SIZE)

of steam per according to the chart above on steam loss through leaking steam traps.

2. At 1000 Btu per lb and 75 percent efficiency in the boiler the BTU loss per year is:

$$\frac{15 \text{ lbs/hr} \times 100 \text{ btu/lb} \times 8760 \text{ hr}}{.75 \text{ eff}} = 175 \text{ mill Btu/yr}$$

3. Savings if the trap are repaired or replaced at $5 per mill Btu of fuel:

    1 trap = $875 per year
    20 traps = $17,500 per year

Guidelines for Selecting Steam Traps
1. Evaluate the traps' basic design, construction material and size of orifice. The most desirable trap is one that is:
    *   Erosion resistant.
    *   Just large enough to carry the condensate load.
    *   Tolerant of rust, corrosion and other particulate from the pipes.
    *   Has a steam loss of less than one pound per hour.

2. Maintain the traps properly by cleaning the trap body at least once a year. Install strainers upstream of each trap to filter the larger particles that could clog the traps. Make sure that the piping system has the proper slope to prevent "water hammer," or slugs of condensate that bolt into the trap and cause premature trap failure or damage.

3. Evaluate the failure mode of the trap, since, depending on the application, it may be important to have the trap fail open or closed. For example, if the process or product dependent on the steam trap is crucial to protect, it may be preferable to have the trap fail open, enabling steam to continue through the lines.

4. Size the steam trap very carefully. If it is oversized, it may leak steam into the return line, resulting in substantial steam loss. If it is undersized, it may cause condensate to back up into the system and lower the operating pressure and temperature. A properly-sized steam trap should not have a steam loss greater than one pound per hour.

5. After installing the proper steam traps establish a program of routine, systematic inspection, testing and trap repair.

Table 15-1
Boiler Combustion Controls

Measuring Stack Temperatures and Oxygen Content
Controlling Combustion Air and Fuel

No. of Boilers	Direct Cost Each	Misc. Matl	Labor Man Hrs	Total Matl & Labor	
				Direct Costs	With 30% O & P
1	3000.00	400.00	20.00	3940.00	5122.00
2	5500.00	800.00	36.00	7272.00	9453.60
3	8000.00	1200.00	52.00	10604.00	13785.20
4	10500.00	1600.00	68.00	13936.00	18116.80
5	13000.00	2000.00	84.00	17268.00	22448.40

Included:
1. Oxygen and temperature sensor
2. Combustion air actuator
3. Load potentiometer
4. Controller

Not Included
1. Extractor
2. Combustibles detector

Direct labor costs are $37.00 per hour.

Figure 15-2

TYPICAL BOILER EFFICIENCY CURVE

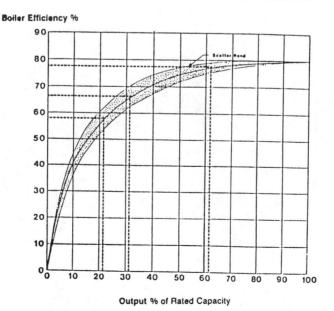

Boiler Efficiency %

Output % of Rated Capacity

Figure 15-3
Combustion Trim Control
For Jackshaft-Controlled Boilers

Figure 15-4
Maximum efficiency can be maintained with this trim control for
jackshaft boilers.

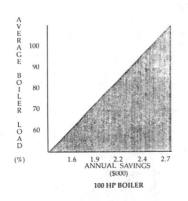

100 HP BOILER

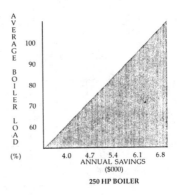

250 HP BOILER

Figure 15-5
Fuel Savings Estimate with Combustion Trim Controls

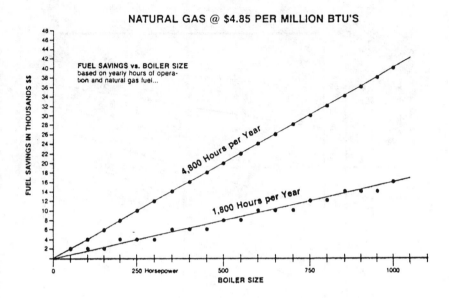

NATURAL GAS @ $4.85 PER MILLION BTU'S

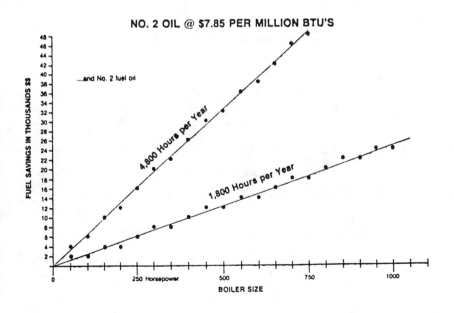

NO. 2 OIL @ $7.85 PER MILLION BTU'S

Figure 15-6
Boiler Combustion Efficiency

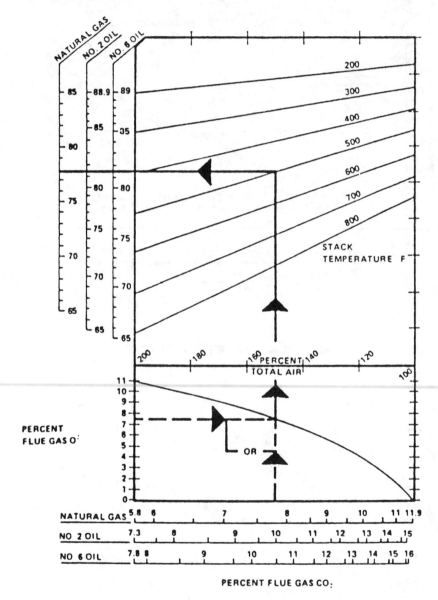

Heating effect of flue gas composition and temperature on boiler efficiency

Table 15-2
Hot Water Reheat Coils

Flanged, 2 Row—Water 200°F in, 180°F out, 3-4 FPS—Air 70°F in, 110°F out, 700 FPM

Btuh	Typical Size	Sq Ft Area	GPM	CFM	Direct Material Cost Each	Per Sq Ft	Labor Man Hours	Total Material & Labor Direct Costs	With 30% O&P
15,100	12×6	0.5	1.5	350	$188	$375.36	1.7	$251	$326
30,200	12×12	1.0	3.0	700	218	218.04	1.8	285	370
45,400	18×12	1.5	4.5	1,050	246	163.76	2.0	320	416
60,500	24×12	2.0	6.1	1,400	270	135.24	2.2	352	457
75,600	24×18	2.5	7.6	1,750	297	118.68	2.4	386	501
90,700	36×12	3.0	9.1	2,100	317	105.80	2.6	414	538
105,800	36×14	3.5	10.6	2,450	345	98.57	2.8	449	583
121,000	36×16	4.0	12.1	2,800	373	93.15	3.0	484	629
151,200	36×20	5.0	15.1	3,500	455	91.08	3.5	585	760
181,400	48×18	6.0	18.1	4,200	552	92.00	4.0	700	910
241,800	48×24	8.0	24.1	5,600	718	89.70	5.0	903	1173

Includes complete installation of coil in duct. Does not include valves or piping connections.
Correction Factors
1. 6 Row Coils, Material add 25%, Labor add 15%.
Direct labor costs are $37.00 per hour.

Table 15-3
Electric Duct Heaters, Slip In
Includes Thermal Cutouts, Fused Transformers, Contractors, Air Flow Switch, Controls, Fan Interlock Temperature Rise 50°F

kW	Btuh	CFM Size Fpm	@500	Amps	Direct Material Costs	Labor Man Hours	Total Material & Labor Direct Costs	With 30% O&P
120 Volt, 1 Phase, Single Stage								
1	3,416	8×6	80	9	$236	1.3	$284	$369
2	6,832	8×8	160	17	258	1.3	306	398
3	10,248	10×8	240	25	265	1.3	313	407
4	13,664	10×10	320	34	273	1.4	325	423
5	17.080	12×10	400	42	286	1.4	337	439
208 Volt, 1 Phase								
1	3,425	8×6	80	5	$251	1.3	$299	$389
2	6,832	8×8	160	10	264	1.3	312	405
3	10,248	10×8	240	15	277	1.3	325	423
4	13,664	10×10	320	20	294	1.4	346	449
5	17,080	12×10	400	24	302	1.4	354	460
7.5	25,620	10×10	600	36	338	1.5	394	512
10	34,160	18×12	800	48	471	1.7	533	694
20	68,320	26×18	1,600	96	644	2.3	730	948
25	85,400	40×18	2,000	120	794	2.8	897	1,166

208 or 240 Volt, 3 Phase

3	10,248	10×8	240	8	$224	1.3	$272	$353
5	17,080	12×10	400	12	311	1.4	362	471
10	34,160	18×12	800	24	404	1.7	467	607
15	51,240	30×12	1,200	36	476	2.0	550	715
20	68,320	26×18	1,600	48	595	2.3	680	884
30	102,480	40×18	2,400	72	755	2.6	851	1,106
50	170,800	48×24	4,000	120	1,043	2.8	1,147	1,491
75	256 200	60×30	6,000	180	1,484	3.8	1,624	2,111
100	341 600	60×36	8,000	240	1,925	5.0	2,110	2,743

480 Volt, 3 Phase

5	17,080	12×10	400	6	$351	1.4	$402	$523
10	34,160	18×12	800	12	457	1.7	520	676
15	51,240	30×12	1,200	18	538	2.0	612	796
20	68,320	26×18	1,600	24	672	2.3	757	984
50	170,800	48×24	4,000	60	1,179	2.8	1,282	1,667
75	256,200	60×30	6,000	90	1,677	3.8	1,817	2,362
100	341,600	60×36	8,000	120	1,826	5.0	2,011	2,614
125	393,250	60×48	10,000	150	1,955	7.0	2,214	2,879
150	512,400	72×48	12,000	180	2,174	9.0	2,507	3,258

Direct labor costs are $37.00 per hour.

Table 15-4
Estimating Duct Heaters
Gas Fired, Indoor

Includes Controls, Burners, Stainless Steel Heat Exchanger, Electric Ignition

Heating Btuh Input	Cfm 70 Output	Rise	Direct Material Cost Each	Labor Per 1,000 Btu	Man Hours	Total Material & Labor Direct Cost	With 30% O&P
25,000	20,000	325	$527	$21.09	4	$675	$878
50,000	40,000	650	607	12.14	5	792	1,030
75,000	60,000	975	662	8.83	5	847	1,102
100,000	80,000	1,300	718	7.18	6	940	1,221
150,000	120,000	2,000	989	6.60	7	1,248	1,623
200,000	160,000	2,600	1,134	5.67	8	1,430	1,859
250,000	200,000	3,300	1,468	5.87	9	1,801	2,342
300,000	240,000	4,000	1,652	5.51	10	2,022	2,628
350,000	280,000	4,600	1,798	5.14	11	2,205	2,867
400,000	320,000	5,300	1,957	4.89	12	2,401	3,121

Table 15-5
Estimating Unit Heaters
Gas Fired, Suspended Indoors
Includes Burners, Gas Valve, Stainless Steel Heat Exchangers, Electric Ignition

Heating Btuh	Cfm 70		Direct Material Cost	Labor	Man Hours	Total Material & Labor	
Input	Output	Rise	Each	Per 1,000 Btu		Direct Cost	With 30% O&P
25,000	20,000	325	$490	$19.60	5	$675	$877
50,000	40,000	650	562	11.23	6	784	1,019
75,000	60,000	975	573	7.64	6	795	1,033
100,000	80,000	1,300	675	6.75	7	934	1,214
150,000	120,000	2,000	865	5.77	8	1,161	1,510
200,000	160,000	2,600	1,045	5.22	9	1,378	1,791
250,000	200,000	3,300	1,289	5.16	10	1,659	2,157
300,000	240,000	4,000	1,524	5.08	11	1,931	2,510
350,000	280,000	4,600	1,801	5.15	12	2,245	2,918
400,000	320,000	5,300	1,976	4.94	13	2,457	3,194

Add for flue, installed $143.16
Add for thermostat & switch $67.81
Add for typical electrical hookup $75.35
Add for typical gas piping, 30 ft. $301.39
Deduct for aluminum heat exchanger 10%
Direct labor costs are $37.00 per hour.

Table 15-6
Gas Fired Cast Iron Boilers
Hot Water and Steam

Heating Btuh			Cfm 70	Direct Material Cost	Labor		Total Material & Labor	
Input	Output	Rise		Each	Per 1,000 Btu	Man Hours	Direct Cost	With 30% O&P
146,000	127,000	110,000		$1,695	$11.61	16	$2,287	$2,973
251,000	218,000	188,000		2,411	9.61	23	3,262	4,240
352,000	306,000	264,000		3,164	8.99	26	4,126	5,364
450,000	391,000	337,000		3,654	8.12	29	4,727	6,145
548,000	477,000	411,000		4,896	8.93	31	6,043	7,856
830,000	722,000	622,000		7,119	8.58	37	8,488	11,035
1,012,000	880,000	759,000		7,535	7.45	46	9,237	12,008
1,394,000	1,212,000	1,056,000		10,398	7.46	50	12,248	15,923
1,624,000	1,412,000	1,248,000		12,206	7.52	54	14,204	18,465
1,980,000	1,722,000	1,537,000		14,617	7.38	65	17,022	22,129

2,320,000	2,017,000	1,801,000	17,180	7.41	74	19,918	25,893
2,784,000	2,421,000	2,162,000	20,194	7.25	82	23,228	30,196
3,092,000	2,689,000	2,400,000	22,152	7.16	88	25,408	33,030
3,710,000	3,226,000	2,880,000	26,071	7.03	95	29,586	38,462
4,330,000	3,765,000	3,361,000	30,592	7.07	112	34,736	45,157
4,950,000	4,304,000	3,843,000	34,057	6.88	122	38,571	50,142
6,180,000	5,374,000	4,800,000	51,237	8.29	144	56,565	73,534

1. Includes valves, controls, insulated jacket.
2. Oil fired hot water or steam versus gas fired, .96 multiplier on material.
3. Add for air eliminator package which includes expansion tank, air vent and fill valve.

110 - 264 MBH $100.96
337 - 759 MBH $111.52
1,056 - MBH & UP $129.60

4. Gross output is the amount of heat needed to heat the building plus piping radiation losses in the distribution system plus pickup allowances for warmups, etc.
5. Net output is the amount needed to heat the building.

Direct labor costs are $37.00 per hour.

Table 15-7
Baseboard Heating
Per Foot

Description	Size	Btuh	Material Cost Per Ft	Labor Man Hours	Total Material & Labor per Ft	
					With Direct	30% O&P
HOT WATER RADIATION (Panel, Cast Iron, with Damper, Fin Tube, Wall Hung, Supports, Excluding Covers)			$23.81	0.8	$53.41	$69.43
ALUMINUM FIN Copper Tube	1-1/4"		30.98	1	67.98	88.38

STEEL FIN					
Steel Tube	1-1/4 "	27.32	1	64.32	83.62
Steel Tube	2 "	31.05	1.3	79.15	102.90
PACKAGE					
ALUMINUM FIN					
Copper Tube	1/2 "	7.59	0.62	30.53	39.69
Copper Tube	3/4 "	7.94	0.67	32.73	42.54
Copper Tube	1 "	13.80	0.67	38.59	50.17
Copper Tube	1-1/4 "	19.67	0.67	44.46	57.79
STEEL FIN					
Iron Pipe Size					
Steel Tube		22.08	0.67	46.87	60.93
CONVECTOR UNIT					
(Damper,Flush,Trim,					
Floor Indented)		36.92	0.8	66.52	86.47

Typical Boiler Efficiency Curve

Table 15-8
Infrared Units
Gas Fired, Electronic Ignition, No Vents,
100% Shut Off

Mbh	Direct Material Cost		Labor	Total Material & Labor	
	Per Each	Man Mbh	Direct Hours	With Cost	30% O&P
15	$380	$25.30	2	$454	$590
30	421	14.03	3	532	691
45	511	11.35	3	622	808
50	524	10.49	4	672	874
60	587	9.78	4	735	955
75	656	8.74	5	841	1093
90	731	8.13	6	953	1239
105	863	8.21	8	1159	1506
120	911	7.59	8	1207	1569

No piping or wiring included

Table 15-9
Electric Baseboard Heating
Commercial Grade, 187W Per Ft, 641 Btu Per Ft

Length Ft	Watts	Btuh	Direct Material Cost		Labor	Direct Costs	Total Material & Labor	
			Each	Per Ft	Man Hours		With 30% O&P	Per Ft
2	375	1,281	$34.50	17.25	1	$71.50	$92.95	$46.48
3	500	1,708	45.54	15.18	1	82.54	107.30	35.77
4	750	2,562	55.20	13.80	1.2	99.60	129.48	32.37
5	935	3,194	75.90	15.18	1.4	127.70	166.01	33.20
6	1,125	3,843	85.56	14.26	1.6	144.76		0.00
7	1,310	4,475	103.50	14.79	1.8	170.10	221.13	31.59
8	1,500	5,124	115.92	14.49	2	189.92	246.90	30.86
9	1,680	5,739	124.20	13.80	2.2	205.60	267.28	29.70
10	1,875	6,405	131.10	13.11	2.4	219.90	285.87	28.59

Table 15-10
Electric Wall Heaters
Commercial

Watts	Btuh	Direct Material Cost Each	Labor Man Hours	Total Material & Labor Direct Costs	Total Material & Labor With 30% O&P
750	2,562	$62.10	1.1	$102.80	$133.64
1,000	3,416	110.40	1.1	151.10	196.43
1,250	4,270	110.40	1.3	158.50	206.05
1,500	5,124	165.60	1.6	224.80	292.24
2,000	6,832	172.50	1.6	231.70	301.21
2,500	8,540	179.40	2	253.40	329.42
3,000	10,248	186.30	2	260.30	338.39
4,000	13,664	200.10	2.3	285.20	370.76

Table 15-11 — Air Curtain Units (Steam and Hydronic Radiation Unit)

Air Curtain Blower Units			Direct Material Cost		Labor	Total Material & Labor	
Mbh	Length	Cfm	Each	Per Cfm	Man Hours	Direct Cost	With 30% O&P
Steam							
200	48" long,	3,980	$624.86	$0.16	16.2	$1,223	$1,590
230	60" long,	4,845	730.85	0.15	16.2	$1,328	$1,727
260	84" long,	3,050	948.34	0.31	20.0	$1,688	$2,195
285	96" long,	3,340	948.34	0.28	22.9	$1,797	$2,336
360	120" long,	4,320	1,126.08	0.26	28.5	$2,181	$2,835
400	144" long,	5,000	1,269.60	0.25	34.0	$2,528	$3,286
530	120" long,	9,570	1,126.08	0.12	29.0	$2,199	$2,859
630	144" long,	11,375	1,269.60	0.11	35.5	$2,583	$3,358
Hydronic							
60	36" long,	1,370	624.86	0.46	6.5	$865	$1,125
75	48" long,	1,714	624.86	0.36	6.5	$865	$1,125
90	60" long,	2,055	692.21	0.34	6.5	$933	$1,213
100	36" long,	2,950	624.86	0.21	6.5	$865	$1,125
105	72" long,	2,550	777.22	0.30	8.1	$1,075	$1,398
130	84" long,	3,120	948.34	0.30	8.7	$1,270	$1,651
150	96" long,	3,420	688. 00	0.20	9.7	$1,046	$1,360
190	120" long,	4,200	1,126.08	0.27	12.9	$1,603	$2,084
230	144" long,	5,110	928.00	0.18	12.9	$1,405	$1,827
260	36" long,	7,700	949.44	0.12	9.7	$1,308	$1,701
300	120" long,	9,810	1,407.60	0.14	11.3	$1,826	$2,373
350	144" long,	11,375	1,587.00	0.14	12.9	$2,065	$2,685

NOTE—Costs include connecting units only and no electrical work

Table 15-12
Heating Equipment
Check Off List

BOILERS
— Cast Iron
— Steel Shell
— Scotch Marine
— Gas Fired
— Oil Fired
— Electric
— Combination Gas/Oil
— Coal, Wood etc.
— Steam

BURNERS
— Gun Type
— Impingement Jet
— Flame Retention Oil

DRAFT CONTROLS
— Barometric
— Vent Dampers
— Induced Draft Fan

COILS
— Hot Water
— Steam
— Electric
— In Air Handling unit
— In Duct
— In Sheet Metal Housing

HVAC CENTRAL UNITS
— HV AHU
— HVAC AHU
— Roof Top
— Make Up Air Units
— Furnaces

SPECIALTIES
— Steam Traps
— Steam Condensate Meter
— Separators (Entrainment
  Eliminator)
— Vacuum Breakers
— Expansion Tank
— Automatic Air Vent
— Daerators
— Water Level Controls
— Water Treatment
— Other Valves (See Piping &
  Valves)

PUMPS
— Centrifugal
— Condensate
— Feed Water
— Smaller Impeller
— Install Smaller Pump
— Install Smaller Motor

FLUES, BREECHINGS
— Flue
— Breeching
— Factory Fabricated Stack

ENERGY SAVING ITEMS
— Check Combustion Efficiency
— Tune Up
— Clean Tubes Surfaces
— Install Permanent Combustion
  Sensors
— Install Automatic Combustion
  Control
— Install Automatic
  Temperature Reset Control

HEATING TERMINAL UNITS

- Fan Coil Units, Cabinets
- Induction units
- Unit Heaters
- Duct Heaters
- Baseboard, Fan Tube
- Baseboard, Radiation
- Radiators
- Infrared Units
- Air Curtain Heaters
- VAV Boxes
- Constant Air Volume Boxes

- Replace Oversize Boiler with Modular Boilers
- Economizer
- Insulate Equipment
- Insulate Piping
- Draft Dampers
- Automatic Vent Dampers
- Electronic Ignitions
- Stack Heat Recovery Unit
- Turbulators, Baffles
- Condensate Heat Recovery Unit
- Automatic Viscosity Controls

Chapter 16

Cooling Equipment

This chapter reviews the methods of saving energy on chillers, cooling towers and in refrigeration systems, plus it provides price and labor tables for replacing cooling equipment.

COOLING ENERGY IMPROVEMENTS

Chillers

- Raise chilled water temperature leaving evaporator 2°F.

- Clean condenser tubes yearly or install automatic tube cleaning system.

- Change one large chiller operating at an inefficient partial load to two or more chillers staged to operate at more efficient full loads in sequence.

- Replace older, inefficient chillers at high C.O.P.'s such as .9, 1.0, 1.1 kW/ton to more efficient chillers in .6, .7 kW/ton range.

- Replace oversized chillers running inefficiently and cycling with new high efficiency models.

- Steam absorption chillers need steam the year round and are less efficient. Two-stage absorption chiller uses 30% to 40% less energy than single-stage uses high-pressure steam. With the use of steam, the cooling tower can be smaller, helping reduce the life-cycle cost of the HVAC system.

- Install double bundle reheat condensers in buildings that have very large interior areas and a lot of heat to recover. Even in cold weather heat can be recaptured.

- Install variable volume chilled water pumping system.

- Use efficient staging. Operate one of several compressors at full load before starting second.

- Use multiple staging with flash intercooling.

- Install controls for multiple chillers so that one chiller can run at full capacity before second is activated.

- Increased heat exchanger surface areas increased efficiency.

Cooling Equipment
- Install variable speed drive on chillers compressors.

- Replace existing compressor and motor only to match reduced load and retain exchangers.

- Install open compressor motors on chillers versus hermetic motors.

- Use thermocycle.

- Use strainer cycle.

- Use outdoor air for cooling in mild weather without running any refrigeration equipment.

- Convert dual duct mixing box, terminal reheat and multi-zone systems to systems which do not require simultaneous heat and cooling thereby being able to shut the cooling system in winter.

Cooling Towers
- Lower cooling tower condenser water temperature returning to chiller.

- Use cooling tower water temperature reset.

- Ceramic cooling tower offers life-cycle benefits of 100% thermal performances for over 25 years uses less electricity and needs less maintenance.

Refrigerant Systems

• Reduce head pressure.

• Adjust suction pressure on DX air conditioning system to raise refrigerant suction temperature 4°F.

• Install hot gas heat exchanger to recovery heat.

• Stage condensers:
  First one on at 60°F outside
  Second one on at 80°F outside

• Obtain free cooling by using an auxiliary heat exchanger connected in parallel to the chilled water supply and return piping between the chiller and the cooling coils. See ASHRAE manual.

• Obtain free cooling by installing water circuit interconnection loop in chilled water supply piping before cooling coils.

CHILLER WATER TEMPERATURE RESET

Two of the parameters which determine how efficiently a chiller operates are the leaving chilled water temperature and the condenser water temperature. Generally these temperatures have fixed setpoints calculated to satisfy the design peak cooling load. Since this peak load occurs very rarely, the chiller often works harder and less efficiently than necessary to meet the immediate cooling load.

By resetting the leaving chilled water temperature upward and/or the entering condenser water temperature downward at low or moderate loads, the operating efficiency of the chiller can be improved without affecting system operation. When leaving chilled water temperature is reset upward, the chilled water valves on the cooling coils will open allowing more chilled water to pass through the coil rather than being bypassed.

Resetting the entering condenser water temperature downward requires more energy to operate the cooling tower, however this additional energy used to run the tower will be more than offset by the amount of energy saved by the chiller.

There are several methods of implementing this type of control:

1. One way is to hold the return chilled water temperature constant, allowing the leaving chilled water temperature to float with the load.

2. Another method is to reset the controllers based on the ambient enthalpy (outdoor air heat content).

3. A third method of control, usually implemented through a computerized building automation system, is to monitor the chilled water valve positions. If any one of the valves is FULL OPEN, then the chilled water temperature cannot be raised any further. If none of the valves are FULL OPEN, then the chilled water temperature can be raised a small increment (say 0.5°F or 1.0°F) . If more than one of the valves is FULL OPEN, then the chilled water temperature must be reset downward a small increment.

The nomographs on the following pages are useful in estimating potential savings from resetting chilled water and condenser water temperatures.

COP at various chilled water temperatures was determined by the relationship:

$$COP = \frac{Btu\ output}{Btu\ input}$$

The change in COP was expressed in terms of the COP at nominal conditions and the plotted as a function of leaving chilled water.

1. Enter the chart on the lower horizontal line at the present leaving chilled water temperature.
2. Proceed vertically upward to the intersection of the type of compressor line.
3. Proceed horizontally left at this intersection and read the percent increase in COP.
4. Repeat this for the maximum average chilled water temperature that can be used.
5. The difference between the two percentages represents the COP percent increase that may be achieved by increasing the chilled water temperature.

COP at various condensing temperatures was determined by the relationship:

$$COP = \frac{Btu\ output}{Btu\ input}$$

Figure 16-1
Cooling effect of chilled water temperature on chiller coefficient of performance - COP

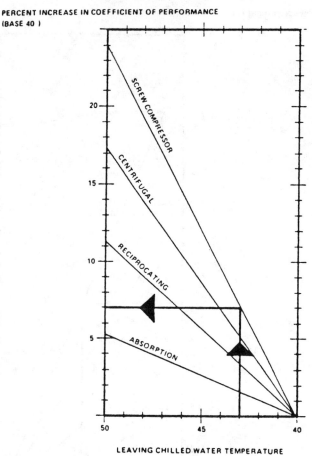

LEAVING CHILLED WATER TEMPERATURE
(°F)

The change in COP was expressed in terms of the COP at nominal conditions and then plotted as a function of condensing temperature.

1. Enter the chart on the lower horizontal line at the amount of reduction in the condensing temperature.
2. Proceed vertically upward to the intersection with the type of compressor line.
3. Proceed horizontally left at this intersection to read the percentage increase in coefficient of performance.

Figure 16-2
Cooling effect of condenser temperature on chiller coefficient of performance - COP

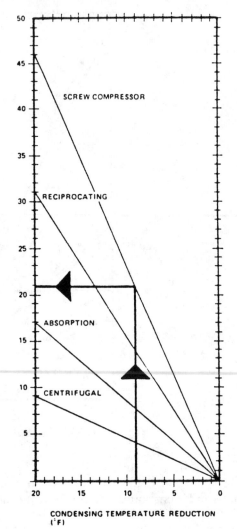

PERCENT INCREASE IN COEFFICIENT OF PERFORMANCE

SCREW COMPRESSOR

RECIPROCATING

ABSORPTION

CENTRIFUGAL

CONDENSING TEMPERATURE REDUCTION
(°F)

Table 16-1
Costs of Clean and Dirty DX Condenser

Nominal Tonnage	kW Hours Per Season		Total Cost Per Season @ 10¢/kW		Extra Costs	Cleaning Cost
	Clean	Dirty	Clean	Dirty	Dirty	
3	4100	5700	410.00	570.00	160.00	156.00
5	5500	8100	550.00	810.00	260.00	156.00
7.5	7400	11200	740.00	1120.00	380.00	168.00
10	12300	16800	1230.00	1680.00	450.00	180.00
15	16000	24400	1600.00	2440.00	840.00	204.00
20	20800	32400	2080.00	3240.00	1160.00	240.00
25	27000	40800	2700.00	4080.00	1380.00	264.00
30	30800	48900	3080.00	4890.00	1810.00	300.00
40	41500	66400	4150.00	6640.00	2490.00	348.00
50	52100	82300	5210.00	8230.00	3020.00	408.00
60	63000	98600	6300.00	9860.00	3560.00	444.00

1. kW Hours based on 1000 hours of operation, 80°F average outside air, and F22 refrigerant.

Figure 16-3
Inlet Condenser Water Temperatures

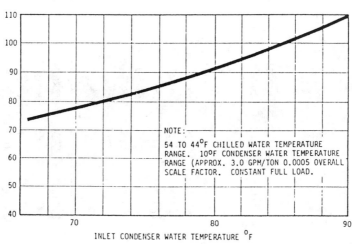

THE EFFECT OF VARIOUS INLET CONDENSER WATER TEMPERATURES ON THE ENERGY CONSUMPTION OF THE COMPRESSOR

Table 16-2
DX Evaporator Coils
6 Row 400 to 600 FPM — Air 75°F in, 55°F out

Btuh	Tons	Typical Size	Sq Ft Area @500 Fpm	Cfm @400 Cfm Per Ton	Direct Material		Labor	Total Material	
					Each	Per Sq Ft	Man Hours	Direct Costs	With 30% O&P
60,000	5	36×16	4	2,000	$529	$132	4	$677	$880
90,000	7.5	36×24	6	3,000	661	110	6	883	1,148
120,000	10	48×24	8	4,000	713	89	7	972	1,264
144,000	12	48×30	9.6	4,80.0	835	87	8	1,131	1,470
180,000	15	48×36	12	6,000	980	82	10	1,350	1,755
240,000	20	48×48	16	8,000	1,081	68	12	1,525	1,982
300,000	25	60×48	20	10,000	1,283	64	14	1,801	2,342
450,000	37.5	72×60	30	15,000	1,639	55	18	2,305	2,997
600,000	50	96×60	40	20,000	2,307	58	22	3,121	4,058
900 000	75	120×72	60	30,000	2,960	49	27	3,959	5,147
1,200 000	100	120×96	80	40 000	3,522	44	32	4,706	6,117
1,500,000	125	120×120	100	50 000	4,250	43	37	5,619	7,305
1,800,000	150	144×120	120	60,000	4,918	41	40	6,398	8,318
2,400,000	200	150×150	160	80,000	6,315	39	48	8,091	10,518

Table 16-3
Chilled Water Coils
6 Row with Drain Pan — Water 45°F in, 55°F out, 3 to 4 FPS — Air 45°F in, 55°F out, 400 to 600 FPM

Btuh	Tons	Typical Size	Sq Ft Area @500 Fpm	Gpm	Cfm @400 Cfm Per Ton	Direct Material Cost Each	Per Sq Ft	Labor Man Hours	Direct Costs	Total Material & Labor With 30% O&P
60,000	5	36×16	4	12	2,000	$480	$120	4	$628	$817
90,000	7.5	36×24	6	18	3,000	600	100	6	822	1,069
120,000	10	48×24	8	24	4,000	649	81	7	908	1,180
144,000	12	48×30	9.6	28.8	4,800	759	79	8	1,055	1,372
180,000	15	48×36	12	36	6,000	890	74	10	1,260	1,638
240,000	20	48×48	16	48	8,000	983	61	12	1,427	1,855
300,000	25	60×48	20	60	10,000	1,166	58	14	1,684	2,189
450,000	37.5	72×60	30	90	15,000	1,490	50	18	2,156	2,803
600,000	50	96×60	40	120	20,000	2,098	52	22	2,912	3,785
900,000	75	120x72	60	180	30,000	2 691	45	27	3,690	4 797
1,200,000	100	120x96	80	240	40,000	3 202	40	32	4,386	5 701
1,500,000	125	120×120	100	300	50,000	3,864	39	37	5,233	6,803
1,800,000	150	144×120	120	360	60,000	4,471	37	40	5,951	7,737
2,400,000	200	150×150	160	480	80,000	5,741	36	48	7,517	9,772

Correction Factors   Material   Labor
1. 2 row coils   .75   .85
2. 4 row coils   .90   .92
Direct labor costs are $37.00 per hour.

Table 16-4
Reciprocating Chillers
Water Cooled With Multiple Semi-Hermatic Compressors

Tons	Direct Material Cost		Labor	Total Material & Labor	
	Each	Per-Ton	Man Hours	Direct Cost	With 30% O&P
20	$14,166	$708	28	$15,202	$19,762
40	21,097	527	32	22,281	28,966
60	28,180	470	36	29,512	38,365
80	35,866	448	40	37,346	48,550
100	43,702	437	48	45,478	59,121
120	51,840	432	56	53,912	70,085
140	57,717	412	62	60,011	78,014
160	64,196	401	66	66,638	86,630
180	69,552	386	70	72,142	93,785

Table 16-5
Reciprocating Chillers
Air Cooled With Compressor

Tons	Each	Per-Ton	Man Hours	Direct Cost	With 30% O&P
20	$15,823	$791	28	$16,859	$21,917
40	23,358	584	32	24,542	31,904
60	31,568	526	36	32,900	42,769
80	40,688	509	40	42,168	54,818
100	47,258	473	48	49,034	63,744
120	59,299	494	56	61,371	79,782

Correction Factors	Materials	Labor
1. Air cooled reciprocating chiller without condenser	0.72	0.75

Labor includes unloading from truck at job site, staging, uncrating, hoisting and setting into place, anchoring, aligning, starting and checkout.
Manhours do not include piping, valves or electrical hookup Based on direct labor costs of $37.00

Table 16-6
Centrifugal Water Cooled Chillers
With Condenser and Single Compressor

Tons	Approx Weight	Direct Material Cost		Labor		Total Material & Labor	
		Each	Per Ton	Man Hours	Direct Cost	With 30% O&P	
100	7,500	$64,046	$640	76	$66,858	$86,915
200	10,000	72,334	362	80	75,294	97,882
300	12,000	85,897	286	88	89,153	115,899
400	18,000	102,473	256	92	105,877	137,640
500	20,000	119,050	238	95	122,565	159,334
600	24,000	137,133	229	98	140,759	182,987
700	24,000	158,231	226	102	162,005	210,606
800	27,000	176,314	220	106	180,236	234,307
900	33,000	195,905	218	112	200,049	260,063
1,000	37,000	213,988	214	118	218,354	283,861
1,200	43,000	253,169	211	124	257,757	335,084
1,400		289,336	207	132	294,220	382,486
1,600		327,010	204	140	332,190	431,847

Labor includes receiving package chiller at job site, unloading, uncrating, setting in place, aligning, etc.
Labor does not include piping, valves or electrical hookup.
Crane rental costs not included.
Chillers with DOUBLE BUNDLE condensers run $80.00 more per ton.
Direct labor costs are $37.00 per hour.

Table 16-7
Cooling Towers

Tons	Gpm	Fan Hp	Direct Material Costs Each	Per Ton	Labor Man Hours	Total Material & Labor Direct Costs	With 30% O&P
Crossflow, induced draft, propeller fan, shipped unassembled*							
100	240	5	$5,658	$57	130	$10,468	$13,608
200	480	7.5	10,764	54	190	17,794	23,132
300	720	15	13,593	45	260	23,213	30,177
400	960	20	15,594	39	330	27,804	36,145
500	1,200	20	18,009	36	370	31,699	41,209
600	1,440	25	20,424	34	400	35,224	45,791
800	1,920	30	25,530	32	450	42,180	54,834
1,000	2,400	35	28,980	29	550	49,330	64,129
Counter flow, forced draft, centrifugal fan, shipped unassembled*							
100	240		$7,728	$77	137	$12,797	$16,636
200	480		14,766	74	200	22,166	28,816
300	720		17,250	58	273	27,351	35,556
400	960		22,770	57	347	35,609	46,292
500	1,200		27,738	55	389	42,131	54,770
600	1,440		32,706	55	420	48,246	62,720
800	1,920		43,470	54	473	60,971	79,262
1,000	2,400		54,510	55	578	75,896	98,665
Induced draft, propeller, galvanized, shipped assembled							
10	24	750 lb	$1,725	$173	12	$2,169	$2,820
15	36	900	2,484	166	13	2,965	3,855
20	48	980	3,174	159	14	3,692	4,800
25	60	1,000	3,795	152	16	4,387	5,703
30	72	1,180	4,347	145	17	4,976	6,469
40	96	1,290	5,520	138	18	6,186	8,042
50	120	1,800	6,072	121	20	6,812	8,856
60	144	2,300	6,293	105	22	7,107	9,239
80	192	3,500	7,066	88	24	7,954	10,340
100	240	4,300	8,280	83	27	9,279	12,063
125	300	4,900	10,005	80	29	11,078	14,401

Table 16-7
Cooling Towers (Continued)

Tons	Gpm	Fan Hp	Direct Material Costs		Labor	Total Material & Labor	
			Each	Per Ton	Man Hours	Direct Costs	With 30% O&P
150	360	6,000	11,592	77	32	12,776	16,609
175	420	6,200	13,041	75	34	14,299	18,589
200	480	6,900	14,352	72	36	15,684	20,389
300	720	9,000	20,700	69	51	22,587	29,363
400	960	14,500	26,496	66	63	28,827	37,475
500	1,200	16,900	31,740	63	72	34,404	44,725

*Redwood, Treated Fir
Man hours include unloading, handling, assembling and set in place. Piping, electrical wiring or crane rental are not included.
Direct labor costs are $37.00 per hour.

Table 16-8
Heat Pumps

Tons of Cooling	Heating Capacity	Direct Material Costs		Labor	Total Material & Labor	
		Each	Per Ton	Man Hours	Direct Costs	With 30% O&P
Split System, Air to Air						
2	8.5	$1,573	$570	9	$1,906	$2,478
3	13	2,167	523	10	2,537	3,298
5	27	3,640	528	20	4,380	5,695
7	33	5,895	610	24	6,783	8,818
10	50	7,735	561	26	8,697	11,306
15	64	10,684	516	30	11,794	15,332
25	119	17,312	502	40	18,792	24,430
30	163	24,909	602	44	26,537	34,498
40	193	27,538	499	50	29,388	38,204

(Continued)

Table 16-8
Heat Pumps (Continued)

Tons of Cooling	Heating Capacity	Direct Material Costs Each	Per Ton	Labor Man Hours	Total Material & Labor Direct Costs	With 30% O&P
Package Unit, Air to Air						
2	6.5	$1,507	$546	5	$1,692	$2,200
3	10	2,167	523	6	2,389	3,105
4	13	2,491	451	8	2,787	3,623
5	27	3,215	466	12	3,659	4,757
7	35	4,524	468	14	5,042	6,554
15	56	11,799	570	18	12,465	16,205
20	100	15,079	546	21	15,856	20,613
25	120	17,961	521	24	18,849	24,503
30	163	22,943	554	26	23,905	31,076
Package Unit, Water Source to Air						
1	13	$1,111	$805	5	$1,296	$1,685
2	19	1,118	405	6	1,340	1,742
3	27	1,511	365	7	1,770	2,301
4	31	1,967	356	8	2,263	2,941
5	29	2,657	385	10	3,027	3,934
7.5	35	3,374	326	16	3,966	5,156
10	50	4,195	304	18	4,861	6,320
15	64	7,335	354	32	8,519	11,074
20	100	8,390	304	36	9,722	12,639

1. Add 5 percent on material for supplementary electric heating coils on air to air units.
2. For water source units add 10 percent to material.
Direct labor costs are $37.00 per hour.

Cooling Equipment
Check Off List

CHILLERS
— Centrifugal
— Reciprocating
— Water Cooled
— Air Cooled
— Package Type
— Remote Condenser
— Heat Recovery Reciprocal
— Double Bundle Condenser

CONDENSERS
— Air Cooled
— Water Cooled

SPECIALTIES
— Receivers
— Valves
    (See Piping and Valves)
— Thermal Expansion Valves

COILS
— Clean Coils
— DX Coils
— Chilled Water Coils
— Number of Rows
— In Air Handling Units
— In Sheet Metal Housing
— Flat Coils
— "A" Coils

COOLING TOWERS
— Packaged Gravity
— Packaged Forced Flow Galvanized
— Crossflow Induced Field Assembled
— Forced Draft Field Assembled

CENTRAL HVAC UNITS
— HVAC Air Handling Units
— Roof Top Units
— Self Contained AC Units
— Air Cooled
— Water Cooled

TERMINAL UNITS
— Terminal Under Window Units Chilled Water
— Induction Units
— VAV Boxes (See VAV Conversions)
— Constant Air Volume Boxes (See Air Handling Equipment)

Figure 16-4
Heat-Pump Worksheet

Would a heat pump reduce your heating bills enough to justify its
cost? Would you be better off buying a new, high-efficiency furnace instead? The worksheet below, used in conjunction with the
table of fuel factors and the climate map, can help you answer

How much heat do you now use?

		CU example	Your house
1.	**Current annual heating bill.** Determine the amount you paid for heating last year. Enter that amount here. If your heating system provides both heat and hot water, enter 85 percent of your bill.	**$1110**	___
2.	**What does the fuel cost?** Enter the price you pay per unit of fuel—gallon of oil, therm of gas, or kilowatt-hour of electricity. If, for example, you pay 7.75 cents per kwh for electricity, enter the price as 0.0775, not 7.75.	**$1.11**	___
3.	**How many fuel units do you use?** Divide line 1 by line 2. Enter the result here.	**1000**	___
4.	**How efficient is the system you have?** Go to the table below and find the fuel factor that's appropriate for the heating system you now have. Enter the factor here.	**98**	___
5.	**How much heat do you use annually?** Multiply line 3 by line 4. Enter the result here. The number you derive is the number of "heat units" (thousands of British thermal units) used to heat the house each year.	**98,000**	___

Fuel factors

Natural gas	60
High-efficiency (condensing) gas furnace	90
Conventional LP gas burner (if LP purchased by the gallon)	55
Heating oil	98
Electric resistance heating	3.4
Water-sourced heat pump	9.4
Air-sourced heat pump (for Zone 1 on map)	8.0
Air-sourced heat pump (for Zone 2)	7.9
Air-sourced heat pump (for Zone 3)	7.5
Air-sourced heat pump (for Zone 4)	6.8
Air-sourced heat pump (for Zone 5)	6.0

CONSUMER REPORTS

those questions. To guide you through the steps, we've included calculations for a homeowner in Bucks County, Pa., who now heats with oil and wants to know if the fuel bill could be cut by installing an air-sourced heat pump.

What would a new heating system cost to run?

	CU *example*	*Your house*
6. *Find the fuel factor.* From the table below, find the fuel factor for the new type of heating system that you're considering. Enter the appropriate figure here. (The factor is 6.8 for our homeowner, who lives in Zone 4 and who is considering an air-sourced heat pump.)	**6.8**	___
7. *How many heat units will the new system use?* Divide line 5 by line 6. Enter the result here.	**14412**	___
8. *What will the fuel cost?* Enter the price you will pay for fuel for the new system: the price per gallon of oil, or therm of natural gas, or kwh of electricity.	**$0.08**	___
9. *What will the system cost to operate?* Multiply line 7 by line 8. Enter the result here.	**$1153**	___

The amount on line 9 is the estimated annual heating bill for the new type of heating system you're considering. By comparing line 9 with line 1, you can determine how much you can expect to save.

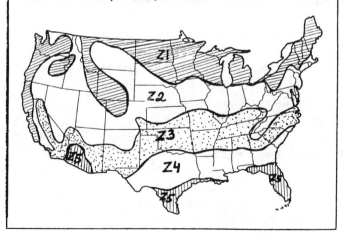

ENERGY RATINGS

Coefficient of Performance (COP)

1. The coefficient of performance is used as a standard of comparison between different refrigeration equipment. The higher the COP, the more beneficial and efficient the refrigeration cycle is. In a refrigeration cycle the COP is the ratio of the cooling effect divided by the input energy expressed in the same units.

$$COP = \frac{Cooling\ Effect}{Energy\ Input}$$

For example the heat absorbed by a DX evaporator, 240,000 Btu (20 tons) divided by the work input of a 20 HP motor into the compressor, which converted to Btu's is, 20 HP x 2500 Btu/HP = 50, 000 Btu.

$$COP = \frac{240,000}{50,000} = 4.8$$

2. The COP for DX vapor compression cycles is normally between 4 and 5.

3. Air to air heat pumps COP's run between 2.8 and 2.9.

4. Improve the efficiency of the cooling equipment if the seasonal COP is:
    Less than 3 for electrical water chillers of compressors
    Less than 2.8 for direct expansion compressors
    Less than 2.5 for self contained HVAC units
    Less than .60 for absorption equipment

Energy Efficiency Ratio (EER)

$$EER = \frac{Btu\ output\ (Cooling\ capacity\ of\ unit)}{Watts\ Input\ (Power\ it\ uses)}$$

$$EER = \frac{12,000\ Btu}{1500\ Watts} = 8$$

The higher the EER the more cooling you get for your dollar. Look for EER'S of 8 or above. An EER of 6 will use 25 percent more energy than an EER of 8 and an EER of 10 will use 25 percent less.

Approximate Power Inputs

System	Compressor kW/ton	Auxiliaries kW/ton
Window Units	1.46	0.32
Through-Wall Units	1.64	0.30
Dwelling Unit, Central Air-Cooled	1.49	0.14
Central, Group or Bldg. Cooling Plants		
(3 to 25 tons) Air-Cooled	1.20	0.20
(25 to 100 tons) Air-Cooled	1.18	0.21
(25 to 100 tons) Water-Cooled	0.94	0.17
(Over 100 tons) Water-Cooled	0.79	0.20

Section 10

HVAC Units and
Air Distribution Equipment

Chapter 17 — Estimating HVAC Units

Chapter 18 — Air Distribution Equipment

Chapter 17

HVAC Units

This chapter lists methods of improving HVAC unit efficiency in order to save electrical and fuel energy and provides price and labor tables for replacing HVAC units.

ESTIMATING HVAC UNITS

Getting more miles per gallon out of HVAC units can reduce energy consumption of electricity and fuels considerably. This can be done by:

1. Using more efficient equipment with high SEER's (Seasonal Energy Efficiency), better C.O.D.'s (Coefficient of Performance), lower kW's per ton with chillers, better mechanical efficiencies of fans and pumps etc.

2. Lowering the resistances in the air and hydronic distribution systems.

3. Improving the efficiencies of heating and cooling equipment by staging them with a series of smaller capacity sections such as burners, heating and cooling coils, chillers etc. to minimize cycling and improving efficiencies as smaller loads.

4. Avoid over sized equipment which over cycles and runs at lower efficiencies.

PRICE AND LABOR TABLES

This chapter contains tables which cover costs and labor of HVAC units such as roof top units, air handling units, self contained air conditioning units, condensing units, makeup air units and room heating and cooling units.

Note that the installation labor shown in the charts is for new construction. For HVAC retrofit projects the time for removing the old equipment is usually half of new installation time. The additional time for installing the replacement equipment for existing building conditions is usually about 35 percent more time than new building installation. These labor adjustments to be factored in to the labor estimates.

ESTIMATING HVAC UNITS

Improve Equipment Efficiency
- High SEER's (Seasonal Energy Efficiency).
- Lower kW's per ton for chillers.
- Higher seasonal C.O.P.'s (Coefficient of Performance).
- Lower equipment pressure drops.
- Keep combustion efficiencies at peaks.
- Keep refrigerants at efficient levels.
- Stage electric heating coils.
- Stage burners.
- Stage compressors.
- Avoid oversized equipment.
- Minimize cycling.

Table 17-1 — Estimating Roof Top Units
Single Zone, DX Cooling Electric Heating Coils
Includes Economizers. Coils, Filters, Curbs, Standard Controls, Warranty

Tons	Cfm	Electric Coil kW	Direct Material Cost Each	Direct Material Cost Per Ton	Direct Material Cost Per Cfm	Labor Man Hours	Total Material & Labor Direct Cost	Total Material & Labor With 30% O&P
2	800	15	$1,912	$956	$2.39	4	$2,060	$2,678
3	1,200	20	2,263	754	1.89	4	2,411	3,135
5	2,000	25	3,594	719	1.80	6	3,816	4,960
7.5	3,000	37	6,405	854	2.14	8	6,701	8,712
10	4,000	50	8,410	841	2.10	8	8,706	11,318
12.5	5,000	62	$10,422	$834	$2.08	10	$10,792	$14,030
15	6,000	75	12,314	821	2.05	10	12,684	16,489
20	8,000	87	16,102	805	2.01	12	16,546	21,509
25	10,000	100	19,767	791	1.98	12	20,211	26,274
30	12,000	125	$23,292	$776	$1.94	16	$23,884	$31,049
40	16,000	150	30,478	762	1.90	20	31,218	40,583
50	20,000	190	37,378	748	1.87	22	38,192	49,650
60	24,000	220	43,129	719	1.80	26	44,091	57,319

CORRECTION FACTORS

		Material	Labor
1. Variable air volume unit		1.28	1.10
2. Multizone units		1.33	1.25
3. Heat Pumps (with electric heating coils)		1.30	
4. Cooling only, no heating		0.66	
5. Power return fan section, add direct costs	10 tons	$1,629	1.10
	20 tons	2,442	1.10
	50 tons	5,373	1.10
6. Omit economizer, deduct	10 tons	$1,058	0.95
	20 tons	1,303	0.95
	50 tons	1,953	0.95

Table 17-2 — Estimating Roof Top Units

Single Zone, DX Cooling, Gas Heating, Staged Cooling and Heating, 7-1/2 Ton and Up
Includes Economizers, Coils, Curbs, Filters, Standard Controls, Warranty

Tons	Cfm	Heating Mbh	Direct Material Cost			Labor	Total Material & Labor	
			Each	Per Ton	Per CFM	Man Hours	Direct Cost	With 30% O&P
2	800	60	$2,185	$1,092	$2.73	6	$2,407	$3,129
3	1,200	94	2,516	839	2.10	8	2,812	3,655
5	2,000	112	3,993	799	2.00	10	4,363	5,672
7.5	3,000	135	7,127	950	2.38	12	7,571	9,843
10	4,000	200	9,345	934	2.34	12	9,789	12,725
12.5	5,000	225	$11,581	$926	$2.32	14	$12,099	$15,728
15	6,000	270	13,680	912	2.28	14	14,198	18,458
20	8,000	360	17,890	894	2.24	16	18,482	24,026
25	10,000	450	21,960	878	2.20	16	22,552	29,318
30	12,000	540	$25,877	$863	$2.16	18	$26,543	$34,505
40	16,000	675	33,864	847	2.12	24	34,752	45,177
50	20,000	810	41,532	831	2.08	26	42,494	55,242
60	24,000	985	48,879	815	2.04	32	50,063	65,081

CORRECTION FACTORS		Material	Labor
1. Variable air volume unit		1.25	1.10
2. Multizone units		1.30	1.25
3. Hot water heating coils		0.90	—
4. Steam heating coil		0.93	1.10
5. Power return fan section, add direct costs	10 tons	$1,536	1.10
	20 tons	2,304	1.10
	50 tons	5,069	
6. Omit economizer, deduct	10 tons	$998	0.95
	20 tons	1,229	0.95
	50 tons	1,843	0.95

Table 17-3
Air Handling Units
DX Coil, Electric Heating Coil
Single Zone, Fan and Coil Section
Isolators, Throwaway Filters, Motor, Drives

Tons	Cfm	Heating Mbh	Direct Material Cost			Labor	Total Matl. & Labor	
			Each	Per Ton	Per Cfm	Man Hours	Direct Cost	With 30% O&P
3	1,200	95	$1,863	$621	$1.55	4	$2,011	$2,614
5	2,000	112	2,908	582	1.45	9	3,241	4,213
7.5	3,000	135	4,090	545	1.36	10	4,460	5,798
10	4,000	200	5,204	520	1.30	12	5,648	7,342
12.5	5,000	225	$6,124	$490	$1.22	14	$6,642	$8,635
15	6,000	270	6,832	455	1.14	16	7,424	9,651
20	8,000	360	8,115	406	1.01	18	8,781	11,415
25	10,000	450	8 841	354	0.88	20	9,581	12,455
30	12,000	540	10 333	344	0.86	22	11,147	14,492
40	16,000	675	$12,630	$316	$0.79	26	$13,592	$17,669
50	20,000	810	14,735	295	0.74	30	15,845	20,599
60	24,000	984	17,224	287	0.72	34	18,482	24,026
80	32,000	1,312	22,046	276	0.69	42	23 600	30,680
100	40,000	1,640	26,411	264	0.66	50	28 261	36,739

Air conditioning only, no electric heating coil.
Direct material multiplier .75

DX Coil, Gas Burners
Single Zone, Fan and Coil Section
Isolators, Throwaway Filters, Motor Drives

Tons	Cfm	Heating Mbh	Direct Material Cost			Labor	Total Matl. & Labor	
			Each	Per Ton	Per Cfm	Man Hours	Direct Cost	With 30% O&P
3	1,200	95	$2 334	$778	$1.94	4	$2,482	$3,226
5	2,000	112	3 641	728	1.82	9	3,974	5,166
7.5	3,000	135	5,121	683	1.71	10	5,491	7,138
10	4,000	200	6,517	652	1.63	12	6,961	9,050
12.5	5,000	225	$7,667	$613	$1.53	14	$8,185	$10,640
15	6,000	270	8,552	570	1.43	16	9,144	11,888
20	8,000	360	10,160	508	1.27	18	10,826	14,074
25	10,000	450	11,070	443	1.11	20	11,810	15,353
30	12,000	540	12,938	431	1.08	22	13,752	17,878
40	16,000	675	$15,813	$395	$0.99	26	$16,775	$21,808
50	20,000	810	18,449	369	0.92	30	19,559	25,427
60	24 000	984	21,564	359	0.90	34	22,822	29,668
80	32 000	1,312	27,602	345	0.86	42	29,156	37,902
100	40,000	1,640	33,066	331	0.83	50	34,916	45,390

Table 17-4
Estimating Air Handling Units
2 Row Water or Steam Heating Coil, 6 Row Chilled Water Coil
Single Zone, Fan and Coil Section
Isolators, Throwaway Filters, Motor, Drives

Tons	Cfm	Heating Mbh	Direct Material Cost			Labor	Total Matl. & Labor	
			Each	Per Ton	Per Cfm	Man Hours	Direct Cost	With 30% O&P
3	1,200	95	$1,554	$518	$1.30	4	$1,702	$2,213
5	2,000	112	2,427	485	1.21	9	2,760	3,588
7.5	3,000	135	3,414	455	1.14	10	3,784	4,920
10	4,000	200	4,344	434	1.09	12	4,788	6,225
12.5	5,000	225	$5,111	$409	$1.02	14	$5,629	$7,317
15	6,000	270	5,702	380	0.95	16	6,294	8,182
20	8,000	360	6,771	339	0.85	18	7,437	9,668
25	10,000	450	7,378	295	0.74	20	8,118	10,554
30	12,000	540	8,625	288	0.72	22	9,439	12,271
40	16,000	675	$10,542	$264	$0.66	26	$11,504	$14,955
50	20,000	810	12,300	246	0.61	30	13,410	17,433
60	24,000	984	14,376	240	0.60	34	15,634	20,325
80	32,000	1,312	18,401	230	0.58	42	19,955	25,941
100	40,000	1,640	22,043	220	0.55	50	23,893	31,061

Correction Factors on all Air Handling Units		Material	Labor
1. Multi-zone unit instead of single zone,			
DX and Electric		1.40	1.25
DX and Gas		1.32	1.25
HW and CHW Coils		1.50	1.25
2. Suspended installation instead of floor mounted		1.25	
3. With filter, mixing box (add)		0.29 /CFM	1.05
4. Variable Air Volume (add)		0.51 /CFM	1.10
5. With economizer section (add)	10 tons	$998	1.05
	20 tons	1,229	1.05
	50 tons	1,843	1.05

Table 17-5
Self-Contained Air Conditioning Units
Air Cooled, DX Coil, Condenser Section,
Electric Heating Coil, Supply Fan, Filters

Tons	Cfm	Heating Mbh	Direct Material Cost			Labor	Total Matl. & Labor	
			Each	Per Ton	Per Cfm	Man Hours	Direct Cost	With 30% O&P
3	1,200	94	$2,995	$998	$2.50	6	$3,217	$4,182
5	2,000	115	3,442	688	1.72	12	3,886	5,051
7.5	3,000	135	4,934	658	1.64	14	5,452	7,088
10	4,000	200	6,325	632	1.58	15	6,880	8,944
12.5	5,000	225	7,787	623	1.56	18	8,453	10,989
15	6,000	270	9,201	613	1.53	20	9,941	12,923
20	8,000	360	12,043	602	1.51	22	12,857	16 714
25	10,000	450	14,177	567	1.42	24	15,065	19 584
30	12,000	540	16,292	543	1.36	26	17,254	22,431
40	16,000	675	20,126	503	1.26	31	21,273	27,655
50	20,000	810	23,401	468	1.17	36	24,733	32,153
60	24,000	985	27,794	463	1.16	42	29,348	38,153

CORRECTION FACTORS   Material   Labor
1. Water cooled
2. Water or steam heating coils.
Direct labor costs are $37.00 per hour.

Table 17-6
Room Heating and Cooling Unit Ventilators
2 Pipe, Single Coil, Controls, Filters, Floor Mounted

Btuh Cooling	Direct Material Cost		Labor	Total Material & Labor	
	Each	Per Btu	Man Hours	Direct Cost	With 30% O&P
6,000	$1,794	$0.299	5.0	$1,979	$2,573
9,000	1,829	0.203	5.5	2,032	2,642
12,000	1,967	0.164	6.1	2,192	2,850
18,000	2,139	0.119	7.0	2,398	3,117
24,000	2,443	0.102	9.3	2,787	3,623
30,000	2,795	0.093	11.5	3,220	4,186
36,000	2,864	0.080	13.2	3,352	4,357
42,000	2,967	0.071	18.0	3,633	4,723
48,000	3,071	0.064	23.2	3,929	5,108
60,000	3,692	0.062	37.0	5,061	6,579

CORRECTION FACTORS
1. 4 pipe, 2 coil, heating and cooling, +14%.
2. For separate electric heating coil, +30%.
3. Ceiling hung unit ventilators, +5%.

Table 17-7
Gas-Fired Makeup Air Units

Cfm	Direct Material Cost		Labor	Total Material & Labor	
	Each	Per Cfm	Man Hours	Direct Cost	With 30% O&P
1,000	$710	$0. 71	6	5872	$1, 134
2,000	1,400	0.70	12	1,724	2,241
4,000	2,800	0.70	15	3,205	4,167
6,000	4,080	0.68	20	4,620	6,006
8,000	5,360	0.67	22	5,954	7,740
10,000	6,600	0.66	24	7,248	9,422
12,000	7,800	0.65	26	8,502	11,053
14,000	8,960	0.64	29	9,743	12,666
16,000	10,240	0.64	31	11,077	14,400
18,000	11,340	0.63	34	12,258	15,935
20,000	12,400	0.62	36	13,372	17,384
24,000	14,640	0.61	37	15,639	20,331
28,000	17,080	0.61	38	18,106	23,538
32,000	19,200	0.60	42	20,334	26,434
36,000	21,600	0.60	46	22,842	29,695
40,000	23,600	0.59	50	24,950	32,435

Chapter 18

Air Distribution Equipment

This chapter contains tables which cover direct material costs, installation and total installed costs of grilles, supply diffusers, dampers, louvers, VAV terminals, VAV terminal retrofit kit, air flow stations, all types of fans, sheaves and inlet vane dampers.

PRICE AND LABOR

Note that the installation labor in the tables is for new construction. For HVAC retrofit projects the time for removing the old equipment is generally one half or less than new installation time. The time for installing replacement equipment in existing building conditions is usually 35 percent more than new building conditions. These adjustments in labor must be factored into HVAC retrofit estimates.

313

Table 18-1
Estimating Multiblade Dampers
Automatic-Control and Manual Multiblade Dampers
Opposed Blade With Frames and Bearings

Size	Semi-Perim	Sq Ft	Direct Material Cost		Labor	Total Matl. & Labor	
	Inches		Each	Per Sq Ft	Man Hours	Direct Cost	With 30% O&P
12×6	18	0.5	$71.88	$143.77	1.0	$109	$142
12×12	24	1.0	75.07	75.07	1.1	116	151
18×12	30	1.5	83.05	55.37	1.3	131	170
18×18	36	2.3	91.85	39.94	1.3	140	182
24×12	36	2.0	84.66	42.33	1.4	136	177
24×24	48	4.0	124.59	31.15	1.8	191	249
30×12	42	2.5	$97.86	$39.14	1.5	$153	$199
30×18	48	3.8	121.40	31.95	1.8	188	244
30×24	54	5.0	137.38	27.48	2.0	211	275
36×18	54	4.5	131.57	29.24	2.0	206	267
36×36	72	9.0	143.19	15.91	2.4	232	302
42×18	60	6.0	143.19	23.86	2.2	225	292
42×24	66	7.0	156.53	22.36	2.4	245	319
48×24	72	8.0	172.51	21.56	2.6	269	349
48×48	96	16.0	330.66	20.67	3.5	460	598
54×18	72	6.8	$151.76	$22.32	2.6	$248	$322
54×24	78	9.0	193.28	21.48	2.8	297	386
60×24	84	10.0	212.45	21.25	3.0	323	420
60×48	108	20.0	408.94	20.45	4.1	561	729
72×24	96	12.0	250.79	20.90	3.5	380	494
72×48	120	24.0	488.80	20.37	4.3	648	842
72×72	144	36.0	661.19	18.37	5.5	865	1,124
84×36	120	21.0	$428.09	$20.39	4.3	$587	$763
84×84	168	49.0	977.61	19.95	6.0	1,200	1,559
96×48	144	32.0	643.74	20.12	5.6	851	1,106
96×96	192	64.0	1277.91	19.97	6.0	1,500	1,950
120×60	180	50.0	998.37	19.97	6.5	1,239	1,611
120×96	216	80.0	1,597.38	19.97	8.5	1,912	2,485

CORRECTION FACTORS	Material	Labor
1. Manual multiblade dampers	0.90	1.00

Table 18-2
Estimating Fire Dampers
Curtain Type Blades, Vertical Installation

Size	Semi-Perim	Sq Ft	Direct Material Cost		Labor	Total Matl. & Labor	
	Inches		Each	Per Sq Ft	Man Hours	Direct Cost	With 30% O&P
12×6	18	0.5	$24.18	$48.36	1.0	$61	$80
12×12	24	1.0	27.16	27.16	1.1	68	88
18×12	30	1.5	35.94	23.96	1.3	84	109
18×18	36	2.3	46.33	20.14	1.3	94	123
24×12	36	2.0	42.19	21.09	1.4	94	122
24×24	48	4.0	70.28	17.57	1.8	137	178
30×12	42	2.5	$48.73	$19.49	1.5	$104	$135
30×18	48	3:8	66.78	17.57	1.8	133	173
30×24	54	5.0	87.86	17.57	2.0	162	210
36×18	54	4.5	75.47	16.77	2.0	149	194
36×36	72	9.0	115.02	12.78	2.4	204	265
42×24	66	7.0	98.89	14.13	2.4	188	244
48×24	72	8.0	106.70	13.34	2.6	203	264
48×48	96	16.0	178.92	11.18	3.5	308	401
54×18	72	6.8	$96.13	$14.14	2.6	$192	$250
54×24	78	9.0	115.02	12.78	2.8	219	284
60×24	84	10.0	124.59	12.46	3.0	236	306
60×48	108	20.0	210.86	10.54	4.1	363	471

CORRECTION FACTORS	Material	Labor
1. U.L. Labels	1.10	—
2. Horizontal installation	1.20	—
3. 22 ga. U.L. sleeve	1.40	1.20
4. 22 ga. U.L. sleeve, free area	1.60	1.25
5. Cap for blades out of air stream	1.25	1.25

6. Multiblade type fire dampers, parallel
   blades Material factor of 1.1 times
   MULTIBLADE CONTROL DAMPER prices.

Table 18-3
Estimating Louvers
Extruded Aluminum, 4 Inches Deep, Fixed Blade, Storm Proof
With Screens, Mill Finish

Size	Semi-Perim	Sq Ft	Direct Material Cost		Labor	Total Matl. & Labor	
			Each	Per Sq Ft	Man Hours	Direct Cost	With 30% O&P
	Inches						
12×6	18	0.5	$45.21	$90.42	1.1	$86	$112
12×12	24	1.0	52.74	52.74	1.2	97	126
18×12	30	1.5	73.47	48.98	1.3	122	158
18×18	36	2.3	100.52	43.70	1.5	156	203
24×12	36	2.0	90.42	45.21	1.5	146	190
24×18	42	3.0	122.06	40.69	2.0	196	;255
30×12	42	2.5	$107.36	$42.95	2.0	$181	$236
30×18	48	3;8	137.43	36.17	2.1	215	280
30×24	54	5.0	158.23	31.65	2.3	243	316
36×18	54	4.5	152.57	33.91	2.3	238	309
36×36	72	9.0	238.02	26.45	2.6	334	434
42×18	60	6.0	180.84	30.14	2.4	270	351
42×24	66	7.0	202.53	28.93	2.6	299	388
48×24	72	8.0	221.23	27.65	2.8	325	422
48×48	96	16.0	377.35	23.58	4.0	525	683
54×18	72	6.8	$194.70	$28.63	2.8	$298	$388
54×24	78	10.0	251.66	25.17	3.3	374	486
60×24	84	10.0	251.66	25.17	3.5	381	496
60×48	108	20.0	455.10	22.75	4.5	622	808
72×24	96	12.0	289.33	24.11	4.0	437	569
72×48	120	24.0	528.04	22.00	5.0	713	927
72×72	144	36.0	786.63	21.85	6.0	1,009	1,311
84×36	120	21.0	$474.69	$22.60	5.0	$660	$858
84×84	168	49.0	1,070.70	21.85	7.0	1,330	1,729
96×48	144	32.0	699.23	21.85	6.0	921	1,198
96×96	192	64.0	1,398.46	21.85	8.0	1,694	2,203
120×60	180	50.0	1,092.55	21.85	7.5	1,370	1,781
120×96	216	80.0	1,748.07	21.85	9.0	2,081	2,705

(Continued)

Labor based on scaffold height installation in basement or on 1st floor or man height installation in penthouse wall.

CORRECTION FACTORS		Material	Labor
1. Galvanized			1.20
2. Finishes:	baked enamel add	2.37/sq ft	
	anodized, add	2.98/sq ft	
	duranodic, add	5.34/sq ft	
	fluoropolymer coating, add	8.22/sq ft	

Table 18-4
Estimating Registers
Return Air Registers, Fixed 45° Vanes,
Opposed Blade Dampers, Commercial Grade

Size	Semi-Perim	Sq Ft	Direct Material Cost		Labor	Total Matl. & Labor	
			Each	Per Sq Ft	Man Hours	Direct Cost	With 30% O&P
	Inches						
12×6	18	0.5	$28.75	$57.49	0.8	$58	$76
12×12	24	1.0	37.55	37.55	0.9	71	92
18×12	30	1.5	50.96	33.98	1.0	88	114
18×18	36	2.3	69.83	30.36	1.1	111	144
24×12	36	2.0	60.69	30.35	1.1	101	132
24×24	48	4.0	115.02	28.76	1.3	163	212
30×12	42	2.5	$76.87	$30.75	1.2	$121	$158
30×18	48	3.8	112.91	29.71	1.3	161	209
30×24	54	5.0	139.78	27.96	1.5	195	254
36×18	54	4.5	127.65	28.37	1.5	183	238
36×36	72	9.0	233.37	25.93	1.7	296	385
42×24	66	7.0	188.98	27.00	1.6	248	323
48×24	72	8.0	210.86	26.36	1.7	274	356
48×48	96	16 0	452.06	28 25	2.6	548	713
54×18	72	6.8	$183.71	$27.02	1.7	$247	$321
54×24	78	9.0	234.35	26.04	2.0	308	401
60×24	84	10.0	270.98	27.10	2.3	356	463
60×48	108	20.0	495.19	24.76	3.0	606	788

(Continued)

Correction Factors-Commercial Grades	Material	Labor
1. Supply registers, single deflection dampers............1.10		1.00
2. Supply registers, double deflection dampers...........1.25		1.00
3. Transfer grille, single deflection, no dampers.........0.63		0.60
4. Relief grille in ceiling, lay in, single deflection, no dampers 0.63		0.40
5. Aluminum construction instead of steel 1.10		1.00
6. Lay in type grilles 1.00		0.50
7. Side wall screw in 1.00		0.65
8. Residential light commercial grade 0.60		0.95

Table 18-5
Ceiling Diffusers
Round, Fixed Pattern, Steel, w/OBD. Commercial Grade

Neck Diameter	Direct Material Cost		Labor	Total Material & Labor	
Inches	Each	Per In. of Dia.	Man Hours	Direct Cost	With 30% O&P
6	$43.70	$7.28	0.7	$69.60	$90.49
8	55.75	6.97	0.8	85.35	110.96
10	66.31	6.63	0.9	99.61	129.49
12	76.85	6.40	1.0	113.85	148.01
14	102.48	7.32	1.1	143.18	186.13
16	131.10	8.19	1.2	175.50	228.15
18	159.74	8.87	1.3	207.84	270.19
20	183.84	9.19	1.5	239.34	311.15
24	220.62	9.19	1.6	279.82	363.77
30	275.78	9.19	1.9	346.08	449.90
36	330.92	9.19	2.2	412.32	536.02

Round, Adjustable Pattern, Steel, w/OBD, Commercial Grade

6	$51.24	$8.54	0.7	$77.14	$100.28
8	64.80	8.10	0.8	94.40	122.73
10	76.85	7.69	0.9	110.15	143.20
12	88.91	7.41	1.0	125.91	163.69
14	119.05	8.50	1.1	159.75	207.68

16	152.20	9.51	1.2	196.60	255.58
18	185.36	10.30	1.3	233.46	303.50
20	213.98	10.70	1.5	269.48	350.33
24	256.79	10.70	1.6	315.99	410.79
30	320.99	10.70	1.9	391.29	508.67
36	385.78	10.72	2.2	467.18	607.33

Rectangular, Adjustable Pattern, Steel, w/OBD, Commercial Grade

6×6	$31.95	—	0.7	$57.85	$75.20
9×9	38.87	—	0.9	72.17	93.83
12×12	48.52	—	1.1	89.22	115.99
15×15	41.15	—	1.2	85.55	111.22
18×18	102.77	—	1.5	158.27	205.75
21×21	124.93	—	1.6	184.13	239.37
24×24	145.42	—	1.7	208.32	270.82

CORRECTION FACTORS	MATERIAL	LABOR
1. Aluminum	1.10	1.00
2. Lay-in diffusers	0.65	0.60

Table 18-6
VAV Terminal Boxes
Cooling Only With Pneumatic or Electric Motor, Controls

Cfm Range	Coil Sq Ft	Direct Material Cost	Labor Man Hours	Material & Labor Direct Cost	With 30% O&P
200-400	—	$312.87	2.0	$387	$503
400-600	—	368.85	2.7	469	609
600-800	—	429.12	3.3	551	717
800-1,000	—	476.49	4.0	624	812
1,000-1,500	—	531.02	4.5	698	907
1,500-2,000	—	595.61	5.0	781	1,015
2,000-3,000	—	645.84	5.5	849	1,104

(Continued)

With Reheat Coils, Pneumatic or Electric Motor and Controls

Cfm Range	Coil Sq Ft	Direct Material Cost	Labor Man Hours	Material & Labor Direct Cost	Material & Labor With 30% O&P
200-400	0.8	$452.09	2.7	$552	$718
400-600	1.0	519.54	2.9	627	815
600-800	1.5	625.75	3.6	759	987
800-1,000	2.0	687.46	4.3	847	1,101
1,000-1,500	2.5	851.07	4.9	1,032	1,342
1,500-2,000	3.0	937.19	5.5	1,141	1,483
2,000-3,000	3.5	1016.12	6.0	1,238	1,610

Figure 18-1
VAV Box Retrofit Kits

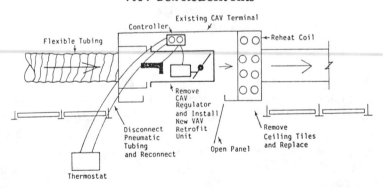

Cost Each

1. For Buensod
    Up to 1000 Cfm $360
2. Titus kits for Titus boxes
    0-500 ....$260
    500-1000 Cfm $360

Figure 18-2
VAV Flow Controller
Add to Existing CAV Box Intake and Disengage CAV on Box

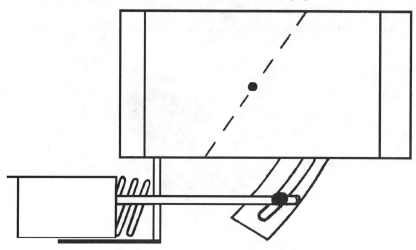

VAV SYSTEMS COMPONENTS

TITUS VAV BOX RETROFIT KITS

		Cost Each

1.   For Buensod or Tutle and Baily CAV boxes
      Up to 1000 Cfm $345

2.   Titus kits for Titus boxes
      0-500 ....$248
      500-1000 Cfm $345

CARRIER VAV UNITS

1.   Moduline cooling only VAV boxes
      with 3 slot diffusers $235

VAV CONTROL PANELS
1.   Controls and monitors supply fans, return fans,
      static pressure sensor, supply and return air, air
      monitoring stations, transmitters, outside air
      control ....$8,280 to $13,800

2. Static pressure regulator $207

Figure 18-3
Slot Cooling Only VAV Box

1. Moduline cooling only VAV boxes
    with 3 slot diffusers $240

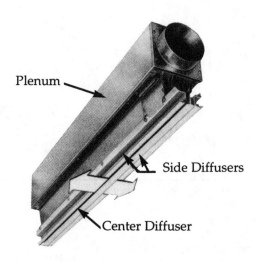

Plenum

Side Diffusers

Center Diffuser

Table 18-7
VAV Ceiling Diffusers

Dia. Inches	Cfm Range	Direct Costs Each	Labor Man hrs	Total Matl & Labor Direct Costs	With 30% O&P
6	100-220	$240	2	$360	$468
8	160-355	$240	2	$360	$468
10	260-580	$240	2	$360	$468
12	380-890	$240	2	$360	$468

Self-contained variable volume ceiling diffusers with thermal sensors for discharge air and warmup, built-in volume controls and built-in pressure sensor for automatic switch-over between cooling and heating.
Direct labor costs are $40 per hour.

Table 18-8
Air Flow Measuring Stations

Size	Sq Ft	Direct Material Costs		Labor	Total Matl & Labor Direct	With 30%
		Each	Per Sq Ft	Man Hr	Costs	O&P
18×12	1.50	108.00	72.00	1.30	143.10	186.03
24×12	2.00	130.00	65.00	1.40	167.80	218.14
24×24	4.00	225.00	56.25	1.60	268.20	348.66
36×24	6.00	300.00	50.00	2.20	359.40	467.22
36×36	9.00	405.00	45.00	2.40	469.80	610.74
48×24	8.00	375.00	46.88	2.60	445.20	578.76
48×36	12.00	480.00	40.00	3.00	561.00	729.30
60×24	10.00	430.00	43.00	3.20	516.40	671.32
60×32	15.00	585.00	39.00	3.50	679.50	883.35
72×36	18.00	684.00	38.00	3.80	786.60	1022.58
72×42	21.00	777.00	37.00	4.30	893.10	1161.03

Direct labor costs are $27.00 per hour.

Air Flow Measuring
Stations   Volumetric Control Center

Figure 18-4
Volumetric Control Center

1. Controls and monitors supply fans, return fans, static pressure
    sensor, supply and return air, air monitoring stations,
    transmitters, outside air control ....$6000 to $10,000
2. Static pressure regulator..$150

Table 18-9
Inlet Vane Dampers
For Centrifugal Fans

Dia. Inches	Area Sq ft	Direct Material Costs Each	Per Sq ft	Labor Man Hrs	Total Matl & Labor Direct Costs	With 30% O&P
18	1.80	70.20	39.00	1.50	109.20	141.96
20	2.20	77.00	35.00	1.70	121.20	157.56
22	2.60	85.28	32.80	1.90	134.68	175.08
24	3.10	94.55	30.50	2.00	146.55	190.52
26	3.70	103.60	28.00	2.20	160.80	209.04
28	4.30	117.18	27.25	2.30	176.98	230.07
30	5.00	128.63	25.73	2.40	191.03	248.34
33	6.00	144.55	24.09	2.60	212.15	275.80
36	7.10	163.30	23.00	2.80	236.10	306.93
39	8.30	182.60	22.00	3.10	263.20	342.16
42	9.60	201.60	21.00	3.30	287.40	373.62
45	11.00	220.00	20.00	3.50	311.00	404.30
48	12.60	239.40	19.00	4.00	343.40	446.42
54	16.00	286.20	17.89	4.60	405.80	527.54
60	19.60	348.69	17.79	5.30	486.49	632.44
66	23.80	411.74	17.30	6.00	567.74	738.06
72	28.20	481.10	17.06	6.50	650.10	845.13
78	33.20	561.74	16.92	7.10	746.34	970.24
84	38.50	650.65	16.90	7.60	848.25	1102.73
90	44.10	742.56	16.84	7.80	945.36	1228.97
96	50.30	840.01	16.70	8.00	1048.01	1362.41

Direct labor costs are $37.00 per hr.

Table 18-10
Estimating Centrifugal Fans
Air Foil Wheel, Single Width Single Inlet, Outlet Velocity 2000 Fpm
Motors, Drives, Isolators Included

Total Costs		Wheel Diam.	Inches Static Press.	Material Cost Total	Per Cfm	Labor To Install Man Hrs	Total Matl & Labor Direct Costs	Sell With 30% O&P
Cfm	Hp							
1,000	1/3	12	1"	$508.56	$.51	3	$589.68	$766.48
2,000	1/2	15	1"	917.28	.47	7	1,103.44	1,434.16
4,000	2	18	1"	1,220.96	.30	10	1,489.28	1,935.44
6,000	2-1/2	22	1-1/2"	1,526.72	.26	12	1,848.08	2,404.48
8,000	3	27	1-1/2"	1,934.40	.25	14	2,308.80	3,002.48
10,000	5	30	2"	2,291.12	.24	16	2,719.60	3,534.96
12,000	10	33	2"	2,694.64	.24	18	3,175.12	4,127.76
14,000	10	33	2"	3,207.36	.24	20	3,742.96	4,865.12
16,000	10	36-1/2	2"	3,463.20	.23	22	4,050.80	5,267.60
18,000	10	40-1/4	2"	3,895.84	.23	24	4,536.48	5,898.88
20,000	10	44-1/4	2"	4,073.68	.21	26	4,769.44	6,199.44
25,000	15	44-1/4	2-1/2"	5,092.88	.21	30	5,894.72	7,662.72
30,000	20	54-1/4	2-1/2"	6,111.04	.21	36	7,074.08	9,196.72
40,000	25	60	2-1/2"	7,638.80	.20	46	8,870.16	11,531.52
50,000	30	66	3"	9,549.28	.20	56	11,046.88	14,360.32
60,000	40	73	3"	11,458.72	.20	64	13,170.56	17,122.56

Table 18-11
Estimating Utility Sets
Forward Curve Wheels, Single Width Single Inlet
Motors, Drives, Isolators Included

Cfm	Hp	Wheel Diam.	Inches Static Press.	Material Cost		Labor To Install Man Hr.	Total Matl & Labor	
				Each	Per Cfm		Direct Costs	Sell With 30% O&P
500	1/4	10	1/2"	$ 253.76	$.51	2	$ 302.64	$ 401.44
1,000	1/2	12	3/4 "	431.60	.44	4	538.72	700.96
2,000	1	12	1"	712.40	.36	5	850.72	1,101.36
4,000	2	18	1"	814.32	.21	7	1,001.52	1,302.08
6,000	3	24	1-1/4"	1,145.04	.20	9	1,386.32	1,802.32
8,000	5	27	1-1/"	1,323.92	.17	10	1,591.20	2,068.56
10,000	5	30	1-1/2"	1,526.72	.16	12	1,848.08	2,404.48
12,000	5	33	1-1/2"	1,832.48	.16	14	2,206.88	2,869.36
14,000	7-1/2	36	1-1/2 "	2,138.24	.16	16	2,566.72	3,336.32
16,000	7-1/2	39	1-1/"	2,240.16	.15	18	2,721.68	3,538.08
18,000	10	40	1-1/2"	2,520.96	.15	20	3,054.48	3,971.76
20,000	15	44	1-1/2"	2,800.72	.15	22	3,389.36	4,405.44

Table 18-12
Industrial Exhaust Fans

Cfm	Direct Material Cost		Labor	Total Matl. & Labor	
	Each	Per Cfm	Man Hours	Direct Cost	With 30% O&P
1,000	$1,130	$1.13	4.00	$1,238	$1,609
2,000	2,080	1.04	9.00	2,323	3,020
4,000	2,640	0.66	13.00	2,991	3,888
6,000	3,420	0.57	16.00	3,852	5,008
8,000	4,400	0.55	18.00	4,886	6,352
10,000	5,300	0.53	21 00	5,867	7,627
12,000	6,360	0.53	23.00	6,981	9,075
14,000	7,420	0.53	26.00	8,122	10,559
16,000	8,160	0.51	29.00	8,943	11,626
18,000	9,180	0.51	31.00	10,017	13,022
20,000	9,200	0.46	34.00	10,118	13,153
24,000	11,040	0.46	39.~0	12,093	15,721
28,000	12,880	0.46	44.00	14,068	18,288
32,000	14,720	0.46	49.00	16,043	20,856
36,000	15,840	0.44	55.00	17,325	22,523
40,000	17,600	0.44	60.00	19,220	24,986

CORRECTION FACTORS
1. For suspended fan, +20%.
2. For DWDI fan, +10%
3. For split centrifugal housing, knocked down, +50%.
4. For inlet vane control, +10%.
5. For crane hoist, with fan set directly on pad, –20%.

Table 18-13

Estimating Roof Exhaust Fans

Centrifugal, Belt Driven, Aluminum Housing, 1/2" Static Pressure
with Shutter, Birdscreen, Curb

Cfm	Hp	Wheel Diam.	Material Cost		Labor To Install Man Hrs	Total Matl & Labor	
			Static Each	Per Cfm		Direct Costs	Sell With 30% O&P
500	1/12	10	$ 449.28	$.90	2	$ 504.40	$ 656.24
1,000	1/6	12	535.60	.53	3	556.40	723.84
1,500	1/4	14	538.72	.42	3	685.36	889.20
2,000	1/3	22	662.48	.34	4	770.64	1,001.52
3,000	1/2	24	749.84	.26	4	855.92	1,114.88
4,000	3/4	24	865.28	.23	5	999.44	1,298.96
6,000	1	30	1,030.64	.18	6	1,191.84	1,548.56
8,000	1-1/2	30	1,220.96	.16	6	1,275.04	1,657.76
10,000	2	36	1,488.24	.16	7	1,675.44	2,177.76
15,000	3	48	2,234.96	.16	8	2,448.16	3,182.40
20,000	5	48	3,077.36	.16	9	3,317.60	4,313.92

Table 18-14
Estimating Vane-Axial Fans
Automatic Controllable Pitch, Direct Drive Includes Inlet Cones, T Frame Motors,
Horizontal Supports, Isolators, Pneumatic Actuator

| Total Costs | | Inches | Material Cost | | Labor To | Total Matl & Labor | |
Cfm	Hp	Wheel Diam.	Static Press.	Total	Per Cfm	Install Man Hrs	Direct Costs	Sell With 30% O&P
Supply Fans Medium Pressure, 1770 RPM								
20,000	30	36/26	5"	$9,115.60	.46	22	$9,726.08	$12,643.28
40,000	50	42/26	5"	10,201.36	.26	32	11,089.52	14,416.48
60,000	100	48/26	6"	12,670.32	.21	44	13,892.32	18,059.60
80,000	150	48/26	6"	14,430.48	.20	48	16,763.76	21,793.20
110,000	200	54/26	6"	17,914.00	.17	64	19,691.36	25,599.60
Return Air Fans, 1170 RPM								
18,000	10	38/26	2"	8,589.36	.48	20	9,144.72	11,888.24
36,000	20	45/26	2"	9,502.48	.27	36	10,500.88	13,653.12
54,000	30	54/26	2"	10,686.00	.20	48	12 019.28	15,624.96
72,000	50	54/26	2"	11,663.60	.17	50	13,052.00	16,771.04
100,000	75	60/26	2"	14,954.16	.16	64	16,732.56	21,753.68

Above fan selection and prices based on Joy vane-axial fans
Direct Labor costs are $37.00 per hour.

Table 18-15
Fixed Pitch Sheaves

Pitch Dia. B Belts	Direct Matl Costs Pulley	Bushing	Labor Man Hrs	Total Matl & Labor Direct Costs	With 30% O&P
FIXED, SINGLE GROOVE, NO BUSHING					
1.9	6.00	—	.2	11.40	14.82
2.3	6.22	—	.2	11.62	15.11
2.6	6.72	—	.3	14.82	19.27
3.0	7.31	—	.3	15.41	20.03
3.2	9.18	—	.4	19.98	25.97
3.4	9.90	—	.4	20.70	26.91
3.9	10.45	—	.4	21.25	27.63
4.6	11.64	—	.5	25.14	32.68
FIXED, DOUBLE GROOVE, BUSHINGS					
3.4	18.30	10.60	.4	39.70	51.61
3.8	19.92	10.60	.4	41.32	53.72
4.6	24.81	10.60	.4	46.21	60.07
4.8	25.47	10.60	.4	46.87	60.93
5.4	28.06	10.60	.4	49.46	64.30
5.8	29.79	10.60	.5	53.89	70.06
6.4	32.36	10.60	.5	56.46	73.40
6.8	33.24	10.60	.5	57.34	74.54
7.0	33.44	10.60	.5	57.54	74.80
8.0	34.27	10.60	.6	61.07	79.39
9.0	35.55	10.60	.6	62.35	81.06
11.0	42.12	10.60	.6	68.92	89.60
12.4	45.49	10.60	.7	74.99	97.49
13.6	54.15	10.60	.7	83.65	108.75
15.4	74.72	10.60	.7	104.22	135.49
18.4	98.94	10.60	.7	128.44	166.97
FIXED, TRIPLE GROOVE, BUSHINGS					
3.4	20.28	10.60	.4	41.68	54.18
3.8	22.54	10.60	.4	43.94	57.12
4.0	23.35	10.60	.4	44.75	58.18
4.4	25.08	10.60	.4	46.48	60.42
4.8	29.79	10.60	.4	51.19	66.55
5.0	30.61	10.60	.5	54.71	71.12
5.4	26.87	10.60	.5	50.97	66.26
5.8	28.69	10.60	.5	52.79	68.63
6.0	29.54	10.60	.5	53.64	69.73

(*Continued*)

Table 18-15 (*Continued*)

Pitch Dia. B Belts	Direct Matl Costs Pulley	Bushing	Labor Man Hrs	Direct Costs	Total Matl & Labor With 30% O&P
6.4	32.05	10.60	.6	58.85	76.51
6.8	33.72	10.60	.6	60.52	78.68
7.0	37.75	10.60	.6	64.55	83.92
8.0	40.28	10.60	.6	67.08	87.20
9.0	45.36	10.60	.6	72.16	93.81
11.0	55.86	10.60	.7	85.36	110.97
12.4	64.54	10.60	.7	94.04	122.25
13.6	69.52	10.60	.7	99.02	128.73
15.4	88.69	10.60	.7	118.19	153.65
18.4	107.42	10.60	.7	136.92	178.00

Direct labor costs are $37.00 per hour.

Table 18-16
Variable Pitch Sheaves

Pitch Dia. B Belts	Direct Matl Costs	Labor Man Hrs	Direct Costs	Total Matl & Labor With 30% O&P
VARIABLE PITCH, SINGLE GROOVE				
2.4-3.2	6.12	.4	16.92	22.00
2.7-3.7	8.08	.4	18.88	24.54
3.1-4.1	9.55	.4	20.35	26.46
3.7-4.7	21.75	.5	35.25	45.83
4.3-5.3	29.40	.5	42.90	55.77
4.9-5.9	35.89	.6	52.09	67.72
5.5-6.5	37.34	.6	53.54	69.60
VARIABLE PITCH, DOUBLE GROOVE				
2.5-3.3	28.45	.5	41.95	54.54
2 9-3.9	32.56	.5	46.06	59.88
3 7-4.7	37.50	.6	53.70	69.81
4.3-5.3	44.20	.6	60.40	78.52
4.9-5.9	56.41	.7	75.31	97.90
5.5-6.5	60.45	.7	79.35	103.16

Table 18-17
V-Belts

Pitch Dia. B Belts	Direct Matl Costs	Labor Man Hrs	Total Matl & Labor Direct Costs	With 30% O&P
33	4.47	.1	7.17	9.32
38	4.89	.1	7.59	9.87
43	5.56	.1	8.26	10.74
47	6.12	.1	8.82	11.47
53	6.75	.2	12.15	15.80
57	7.05	.2	12.45	16.19
63	7.32	.2	12.72	16.54
67	7.80	.2	13.20	17.16
73	8.27	.3	16.37	21.28
76	8.47	.3	16.57	21.54
83	9.42	.3	17.52	22.78
86	9.85	.3	17.95	23.34
93	10.45	.4	21.25	27.63
96	10.92	.4	21.72	28.24
103	11.67	.4	22.47	29.21
106	11.95	.4	22.75	29.58
108	12.24	.5	25.74	33.46
111	12.65	.5	26.15	34.00
115	13.23	.5	26.73	34.75
123	13.88	.5	27.38	35.59
131	14.80	.6	31.00	40.30
139	15.90	.6	32.10	41.73
147	16.83	.6	33.03	42.94
161	18.29	.6	34.49	44.84
176	19.96	.6	36.16	47.01

Figure 18-5
Air Distribution Equipment

Check Off List

FANS
— Centrifugal
— Utility Sets
— Adjustable Pitch Vane Axial
— Vane Axial
— Tubular Centrifugal Propeller
  Roof Exhaust Fans
— Inlet Vane Dampers
— Vibration Isolators
— Access Doors

ROOF HOODS
— Gravity Vents
— OA Hoods
— Louvered Penthouses
— Roof Curbs

AIR DIFFUSION EQUIPMENT
— Grilles, Registers
— Ceiling Diffusers
— Linear Diffusers
— Slot Diffusers
— VAV Ceiling Diffusers
— Light/Air Troffers
— Extractors

TERMINAL EQUIPMENT
— VAV Terminals (See VAV
  Conversions)
— Constant Air Volume
  Terminals

FILTERS
— Fiberglass Throwaway
— Automatic Roll
— Bag Filters
— Medium Efficiency Pleated
— High Efficiency Absolute
  Charcoal Metal Mesh
— Electronic Filters

— DAMPERS
— Multiblade
— Fire Dampers
— Inlet Vane
— Louvers

Section 11

Ductwork and Piping

Chapter 19 -
Ductwork

Chapter 20 -
Pumps, Piping, and Insulation

Chapter 19

Estimating Ductwork

The first two tables in this chapter covers installed budget per square foot prices of all types of ductwork plus per lb and per ft installed prices of low pressure HVAC galvanized ductwork. The prices are for average size ductwork in a typical mix of straight duct and fittings.

The budget prices are based on union shops and the labor portion of the total installed price on new building conditions. The third table gives installed budget prices for HVAC galvanized ductwork for different average gauges and percentage fittings following by a similar graph for labor.

Factors for adjusting to medium and high pressure ductwork, weight tables per foot for rectangular and round galvanized ductwork, square foot per foot of fiber glass ductboard and weight of aluminum ductwork per foot are all included.

Labor tables for spiral pipe and fittings and black iron ductwork are included as well as tables on estimating liner and duct wrap.

Figure 1. Installed Price Per Square Foot of
Different Types of Ductwork
Average Size 24"×12" in a Typical Mix
25 Percent Fittings by Square Feet

HVAC Ductwork
Spiral $4.23
Fiber Glass Dct $4.41
Lp Galv, Bare $5.31
Lp Galv, Insulated $7.04
Mp Galv, Bare $6.53
Mp Galv, Insulated ...$8.46

Industrial Ductwork
Aluminum, Light Ga. $5.50
Pvc Coated Galv, Light Ga. $8.33
Stainless Steel Light Ga. $8.83
Black Iron, Angle Flanges, 16 Ga. $11.08
Black Iron, Angle Flanges, 14 Ga. $12.32
Pvc Plastic $12.40
Frp, Fiber Glass Reinforced Plastic $14.49
Black Iron, Angle Flanges, 10 Ga. $16.44

Installed price per square foot includes material, shop labor, field labor, shop drawings, shipping and a 35 percent markup on costs for overhead and profit. Labor is based on $37.00 per hour.

337

Figure 2. Budget Estimating Galvanized Ductwork
Per Pound and Per Foot
Standard Low Pressure HVAC Rectangular Galvanized
25 Percent Fittings, New Construction, 10 Foot High, 1st Floor

Size	Semi-Perim Inches	Gauge	Lb/Ft w/20% Waste	Sq Ft/Ft No Waste	Selling Price* Furnished & Installed Per Lb.	Per Ft.
6×6	12	26 Ga.	2.8	2.0	$4.79	$13.41
12×6	18		3.3.	3.0	4.62	15.24
12×12	24		4.4	4.0	4.44	20.58
18×6	24	24 Ga.	5.6	4.5	4.44	24.91
18×12	30		7.0	5.0	4.31	30.99
24×9	33		7.7	6.5	4.24	32.58
24×12	36		8.4	6.0	4.17	34.96
24×15	39		9.2	6.5	4.17	38.30
30×12	42		9.8	7.0	4.08	40.09
30×18	48		11.2	8.0	4.02	45.00
30×24	54		12.6	9.0	3.95	49.74
36×12	48	22 Ga.	13.6	8.0	3.95	53.68
36×18	54		15.3	9.0	3.85	58.84
36×24	60		17.0	10.0	3.73	63.44
42×12	54		15.3	9.0	3.85	58.84
42×18	60		17.0	10.0	3.73	63.44
42×24	66		18.7	11.0	3.70	69.25
48×12	60		17.0	10.0	3.73	63.44
48×18	66		18.7	11.0	3.70	69.25
48×24	72		20.4	12.0	3.67	74.98
54×24	78		22.1	13.0	3.64	80.56
54×30	84		23.8	14.0	3.62	86.08
54×36	90		25.5	15.0	3.59	91.49
60×18	78	20 Ga.	26.0	13.0	3.62	94.03
60×24	84		28.0	14.0	7.92	101.26
60×30	90		30.0	15.0	3.59	107.64
72×24	96		32.0	16.0	3.56	113.89

(Continued)

Figure 2. (*Continued*)

Size	Semi-Perim ——— Inches	Gauge	Lb/Ft w/20% Waste	Sq Ft/Ft No Waste	Selling Price* Furnished & Installed	
					Per Lb.	Per Ft.
72×30	102		34.0	17.0	3.53	120.05
72×36	108		35.3	18.0	3.49	123.11
84×30	114		38.0	19.0	3.56	130.89
84×36	120		40.0	20.0	3.42	136.63
84×42	126		42.0	21.0	3.38	142.25
96×24	120	18 Ga.	52.6	21.0	3.35	176.64
96×36	132		57.2	22.0	3.33	190.45
96×48	144		62.4	24.0	3.30	205.98
96×72	168		72.8	28.0	3.27	238.24
96×96	192		83.2	32.0	3.24	269.82

*Selling price based on $.42 per pound for galvanized, $37.00 per hour for labor, a 30 percent markup on direct costs for overhead and a 5 percent markup for profit. The price including material and all labor for the fabrication, installation, drafting and shipping.

Correction factors on selling price of galvanized ductwork.
1. Percent Fittings by Weight,   15% ....0.09   25% 1.00
                                                       35% ....1.08   45% 1.15
2. Different wage rates (incl. base pay, fringes, payroll taxes, ins.)
                              $12.00 (per hour).....0.66 $22.00 ....0.89
                              15.00 ....0.77   26.50 ....1.00
                              18.00 ....0.80   30.00 ....1.08
3. overall residential factor for 15 percent fittings, $12.00 per hour gross wages and lighter gauges...0.50 times cost per lb or per foot.
Example: 24" × 8 $4.15/lb equals $2.08/lb
                         $31-90/ft equals $15.95/ft

Figure 19-3
Installed Cost per Pound for Galvanized Ductwork

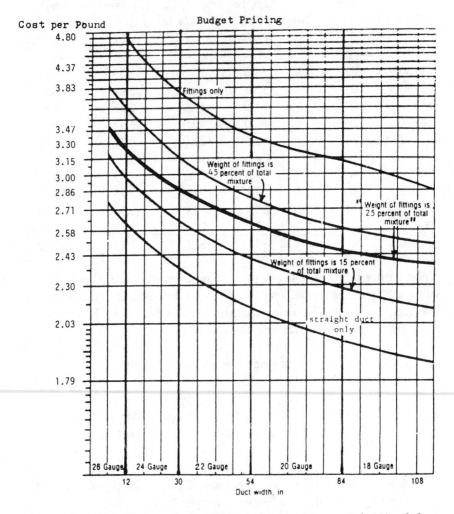

Curves for cost per pound of uninsulated low pressure galvanized ductwork for new construction, 2000 lbs and up, standard installation conditions and conventional duct fabrication (not coil line).

Prices include material, shop and field labor, shop drawings, shipping
and a 30 percent markup for overhead and profit on both material and
labor.

Costs are based on galvanized material at $.35 per pound and direct labor
wage rates as of June 1, 1985 of $27.50 per hour which includes base pay,
fringes, insurance and payroll taxes.

Figure 19-4
Pounds per Hour Fabrication Labor for
Low Pressure Galvanized Ductwork

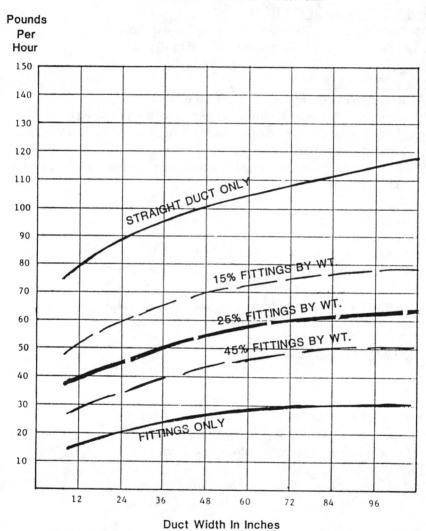

Duct Width In Inches

Figure 19-5
Medium and High Pressure Ductwork

PRESSURE CLASSIFICATIONS

Medium pressure	2000 fpm and up	2 to 6 inches S.P.
High pressure	2000 fpm and up	6 to 10 inches S.P.

CONSTRUCTION

About 2 gauges heavier than low pressure galvanized.

More reinforcing angles required combined with alternate tie rods.

Angle reinforced double S cleats, welded flanges or companion angles are used at connections.

Seams and connections are sealed to withstand pressures 25% over design S.P.

CORRECTION FACTORS

(On the totally installed cost of low pressure if figured as the identical duct system)

Medium pressure cleat connection 1.25

Medium pressure companion angles 1.30

High pressure cleat connection 1.35

High pressure companion angles 1.40

TAKE OFF PROCEDURE

Take off the same as low pressure ductwork, either by the piece or per pound.

METHODS OF PRICING

Completely price all labor and material as low pressure and apply a correction factor from above onto the total-price.

Example:

As Low Pressure	As Medium Pressure Cleats
$30,000	1.25 × $30,000 = $37,500

Or apply factors to the low pressure material and labor separately:

As Low Pressure	As Medium Pressure Angles		
1500 lbs	1.30 ×	1500 =	1950 lbs
35 hrs Shop	1.30 ×	35 =	46 hrs
60 hrs Field	1.30 ×	60 =	78 hrs

BUDGET PRICING (Based on 10,000 lb job, 25% fitt. by wt., standard conditions)

Medium pressure cleat connections	$3.61
Medium pressure companion angles	3.74
High pressure cleat connections	3.89
High pressure companion angles	4.05

Galv. 35¢/lb, labor $27.50/hr. Includes material, all labor cartage, drafting, overhead and profit.

Figure 19-6
Weight of Rectangular Ductwork
Per Linear Foot with 20% Allowance

Semi-Perim Width + Depth	0-12 26 ga .91 lbs/SF 1.10 W/20%	13-30 24 ga 1.16 1.40	31-54 22 ga 1.41 1.70	55-84 20 ga 1.66 2.00	85 up 18 ga 2.16 2.60	Square Feet (no allow.)
12	2.20	2.80	3.40	4.00	5.2	2.00
14	2.58	3.28	3.98	4.68	6.09	2.34
16	2.94	3.74	4.54	5.34	6.95	2.67
18	3.30	4.30	4.2	5.10	6.00	3.00
20	3.68	4.68	5.68	6.68	8.69	3.34
22	4.04	5.14	6.24	7.34	9.55	3.67
24	4.40	5.60	6.80	8.00	10.4	4.00
26		6.08	7.38	8.68	11.29	4.34
28		6.54	7.94	9.34	12.15	4.67
30		7.02	8.50	10.00	13.00	5.00
32		7.48	9.08	10.68	13.85	5.34
34		7.94	9.68	11.34	14.75	5.67
36		8.40	10.10	12.00	15.6?	6.00
38		8.88	10.88	12.68	16.45	6.34
40		9.34	11.32	13.34	17.34	6.67
42		9.8?	11.90	14.00	18.20	7.00
44		10.28	12.48	14.68	19.24	7.34
46		10.74	13.00	15.34	20.00	7.67
48		11.20	13.60	16.00	20.80	8.00
50		11 68	14 27	16 68	21 69	8 34

(Continued)

Figure 19-6 (*Continued*)

Semi-Perim Width + Depth	0-12 26 ga .91 lbs/SF 1.10 W/20%	13-30 24 ga 1.16 1.40	31-54 22 ga 1.41 1.70	55-84 20 ga 1.66 2.00	85 up 18 ga 2.16 2.60	Square Feet (no allow.)
52		12.14	14.93	17.34	22.55	8.67
54		12.6?	15.30	18.00	23.40	9.00
56		13.00	15.87	18.68	24.29	9.34
58		13.54	16.43	19.34	25.15	9.67
60		14.00	17.00	20.00	26.00	10.00
62			17.57	20.68	26.87	10.34
64			18.13	21.34	27.73	10.67
66			18.70	22.00	28.6?	11.00
68			19.27	22.68	29.47	11.34
70			19.83	23.34	30.33	11.67
72			20.40	24.00	31.20	12.00
74			20.98	24.68	32.07	12.34
76			21.55	25.34	32.93	12.67
78			22.10	26.00	33.80	13.00
80			22 67	26.68	34.67	13.34
82			23.23	27.34	35.53	13.67
84			23.80	28.00	36.40	14.00
86			24.37	28.68	37.27	14.34
88			24.93	29.34	38.13	14.67
90			25.50	30.00	39.00	15.00
92			26.07	30.68	40.00	15.34
94			26.63	31.34	40.87	14.67
96			27.20	32.00	41.60	16.00
98			27.77	32.68	42.47	16.34
100			28.33	33.34	43.33	16.67
102			28.90	34.00	44.20	17.00
104			29.47	34.68	45.09	17.34
106			30.03	35.34	46.96	17 67

(*Continued*)

Figure 19-6 (*Continued*)

Semi-Perim Width + Depth	0-12 26 ga .91 lbs/SF 1.10 W/20%	13-30 24 ga 1.16 1.40	31-54 22 ga 1.41 1.70	55-84 20 ga 1.66 2.00	85 up 18 ga 2.16 2.60	Square Feet (no allow.)
108			30.60	36.00	46.80	18 00
110				36.68	47.67	18.34
112				37.34	48.53	18.67
114				38.00	49.40	19.00
116				38.68	50.27	19.34
118				39.34	51.13	19.67
120				40.00	52.00	20.00

Figure 19-7. Spiral Pipe and Fittings
Fabrication Labor—Pipe Feet per Hour; Fittings Hours per Piece

DIA.	PIPE HR/FT	FT/HR	90° ELL 5 Gores	45°,60° ELBOWS 3 Gores	45°or 30° Lateral Straight	Reducing	90° Tees Straight	Reducing	SQ. to RND.	WELD ANGLE RINGS ON EA.
4	.0078	129	1.6	.80	1.8	3.0	1.7	2.7	2.8	.50
5	.0081	124	1.8	.96	1.8	3.0	1.7	2.7	2.9	.51
6	.0085	118	2.0	1.0	1.8	3.0	1.7	2.7	3.0	.52
7	.0089	112	2.1	1.8	1.9	3.0	1.8	2.8	3.1	.54
8	.0093	107	2.4	1.2	1.9	3.0	1.8	2.8	3.2	.55
9	.0099	101	2.6	1.4	2.1	3.3	1.8	3.0	3.4	.57
10	.0105	95	2.8	1.4	2.1	3.3	1.8	3.0	3.4	.58
11	.0111	90	3.0	1.5	2.2	3.4	1.9	3.0	3.7	.59
12	.0119	84	3.2	1.6	2.3	3.7	2.0	3.4	3.8	.60
14	.0135	74	3.8	1.9	2.6	4.1	2.3	3.7	4.0	.64
16	.0154	65	4.3	2.2	2.7	4.3	2.4	4.0	4.2	.69
18	.0175	57	4.8	2.4	3.0	4.9	2.7	4.3	4.2	.74
20	.0192	52	5.6	2.8	3.2	5.1	2.9	4.6	4.6	.78
22	.0213	47	6.4	3.2	3.4	5.5	3.1	5.0	4.7	.86
24	.0233	43	7.2	3.6	3.8	6.0	3.3	5.3	5.0	.95
26	.0244	41	8.0	4.0	4.2	6.8	3.8	6.2	5.3	1.01
28	.0263	38	9.0	4.4	4.6	7.4	4.2	6.6	5.4	1.06
30	.0286	35	9.6	4.8	5.1	8.2	4.6	7.4	5.6	1.12
36	.0345	29	11.2	5.6	6.8	10.9	6.2	9.8	6.4	1.34
42	.0400	25	12.8	6.4	7.8	12.4	7.0	11.1	7.2	1.52
48	.0455	22	14.4	10.2	9.0	14.5	7.8	12.6	8.0	1.75
54	.0526	19	15.2	7.6	10.0	16.1	8.7	13.8	8.8	1.98
60	.0588	17	16.0	8.0	11.0	17.8	9.6	15.1	9.6	2.21

BUDGET FIGURES TO PURCHASE ONLY (Includes mat., fab, and O & P)

```
1.  Spiral Pipe & Fittings .......................   $ .88/LB
    Pipe Only, 420 lb/hr (360 sq ft/hr)............     .58/LB
    Fittings Only, 20.5 lb/hr (21.5 sq ft/hr).......    2.71/LB
    Double Skin Fittings ..........................     3.47/LB

2.  Residential Furnace Pipe.....................   $ .88/LB
```

SPIRAL ACCESSORIES (4% approx., average size 18" dia.)

```
Cement: 15 connections per gallon (105 ft of pipe/gal) @ $17/gal.
Tape:   25 connections per roll (175 ft duct per roll)
        2 inch wide, 180 ft/roll @ $11/roll = 6.3¢/ft.
Shrink bands: $1.98 per ft in rolls
Hangers:  Add 6%
```

Figure 19-8

Weight of Round Steel Ductwork Per Linear Foot with 20% Allowance

Dia. Inches	Square Foot Per Lf	26 ga .75 .90	24 ga 1.00 1.20	22 ga 1.25 1.50	20 ga 1.50 1.80	18 ga 2.00 2.40	16 ga 2.50 3.00	14 ga 3.13 3.76	12 ga 4.37 5.24	10 ga 5.63 lbs/SF 6.76w/20%
					Lbs/Lf					
4	1.05	.95	1.3	1.6	1.9	2.5	3.2	3.9	5.5	7.1
5	1.31	1.2	1.6	2.0	2.4	3.1	3.9	4.9	6.9	8.9
6	1.57	1.4	1.9	2.4	2.8	3.8	4.7	5.9	8.2	10.6
7	1.83	1.6	2.2	2.7	3.3	4.4	5.5	6.9	9.6	12.4
8	2.09	1.9	2.5	3.1	3.8	5.0	6.3	7.9	11.0	14.1
9	2.36	2.1	2.8	3.5	4.2	5.7	7.1	8.9	12.4	16.0
10	2.62	2.4	3.1	3.9	4.7	6.3	7.9	9.9	13.6	17.7
11	2.88	2.6	3.5	4.3	5.2	6.9	8.6	10.8	15.1	19.5
12	3.14	2.8	3.8	4.7	5.7	7.5	9.4	11.7	16.5	21.2
14	3.66	3.3	4.4	5.5	6.6	8.8	11.0	13.8	19.2	24.8
16	4.19	3.8	5.0	6.3	7.5	10.1	12.6	15.8	22.0	28.3
18	4.70	4.2	5.6	7.1	8.5	13	14.1	17.7	24.6	31.8
20	5.23	4.7	6.3	7.8	9.4	12.6	15.7	19.7	27.4	35.4
22	5.76	5.2	6.9	8.6	10.4	13.8	17.3	21.7	30.2	39.0
24	6.28	5.7	7.5	9.4	11.3	15.1	18.8	23.6	32.9	42.5
30	7.85	7.1	9.4	11.8	14.1	18.8	23.6	29.5	41.1	53.1
36	9.42	8.5	11.3	14.1	17.0	22.6	28.3	35.4	49.4	63.7
42	10.99	9.9	13.2	16.5	19.8	26.4	33.0	41.4	57.6	74.4
48	12.56	11.3	15.1	18.8	22.6	30.1	37.7	47.2	65.8	84.9

Figure 19-9. Spiral Pipe and Fittings
Installation Labor

Dia In	Pipe Based On 10 Ft Lgts Hr/Ft	Ft/Hr	Fitting Labor-Per Piece 90° Rad Ells 90° Tees 45° Laterals Wye Fit	Sq. Ells 30° 45° 60° Rad Ells Rect to Rnd
4	.06	16.7	.35	.35
5	.07	14.3	.40	.40
6	.08	12.5	.45	.45
7	.08	12.5	.50	.50
8	.09	11.0	.55	.55
9	.10	10.0	.60	.60
10	.11	9.1	.65	.65
11	.12	8.3	.70	.10
12	.13	7.7	.80	.75
14	.14	7.0	.91	.80
16	.15	6.7	1.10	.90
18	.17	5.9	1.30	1.10
20	.18	5.6	1.40	1.20
22	.20	5.0	1.60	1.25
24	.22	4.5	1.70	1.30
26	.24	4.2	1.90	1.40
28	.26	3.8	2.00	1.50
30	.29	3.5	2.30	1.60
36	.33	3.0	2.90	1.90
42	.39	2.6	3.80	2.60
48	.42	2.4	4.20	3.30
54	.46	2.2	4.40	3.70
60	.50	2.0	5.50	4.10

Correction Factors (Applied to per piece field labor)
Spiral low pressure 0.90
Spiral double skin 1.75
Oval 1.5
Oval double skin 2.0
Underground 1.5
Thermal shrink bands on spiral 0.8

Figure 19-10
Square Foot Per Running Foot of Fiber Glass Ductwork
With Allowances Included

Semi-Perim I.D. Width plus Depth	Sf/Lf Includes allowances 8" corner overlap 5% waste	Sf/Lf No Allowance Included
12	2.8	2.00
14	3.1	2 34
16	3.5	2.67
18	3.9	3.00
20	4.2	3.34
22	4.6	3.67
24	4.9	4.00
26	5.2	4.34
28	5.6	4.67
30	5.9	5.00
32	6.3	5.34
34	7.7	5.67
36	7.0	6.00
38	7.3	6.34
40	7.7	6.67
42	8.0	7.00
44	8.4	7.34
46	8.8	7.67
48	9.1	8.00
50	9.5	8.34
52	9.8	8.67
54	10.1	9.00
56	10.5	9.34
58	10.9	9.67
60	11.2	10.00

(*Continued*)

Figure 19-10 (*Continued*)

Semi-Perim I.D. Width plus Depth	Sf/Lf Includes allowances 8" corner overlap 5% waste	Sf/Lf No Allowance Included
62	11.6	10.34
64	11.9	10.67
66	12.3	11.00
68	12.6	11.34
70	12.9	11.67
72	13.3	12.00
74	13.7	12.34
76	14.0	12.67
78	14.4	13.00
80	14.7	13.34
82	15.1	13.67
84	15.4	14.00
86	15.8	14.34
88	16.1	14.67
90	16.5	15.00
92	16.8	15.34
94	17.2	15.67
96	17.5	16.00
98	17.9	16.34
100	18.2	16.67

Figure 19-11
Calculating Fiber Glass Ductboard Labor

Following labor productivity rates are based on:

1. Volume of 1000 SF and up
2. 800 Board
3. Average duct size of 18"×12", or equivalent 24 gauge galv.
4. Tie rod reinforcing
5. Rates based on gross square footage including 8" overlap and 5% waste
6. Using standard grooving machine.

Fabrication Labor
Size Ranges

Percentage Fittings By Weight	0-12"		13-30"		31-54"		55-84"		85-96"	
	Sf/Hr	Hr/Sf	Sf/Hr	Hr/Sf	Sf/Hr	Hr/Sf	Sf/Hr	Hr/Sf	Sf/Hr	Hr/Sf
Str. Duct Only	59	.0169	62	.0161	56	.0179	53	.0189	38	.0263
10-20%	55	.0183	58	.0174	52	.0191	49	.0202	35	.0284
20-30%	52	.0191	55	.0182	50	.0200	47	.0211	34	.0297
30-40%	50	.0200	53	.0190	48	.0208	45	.0220	32	.0311
Fittings Only	39	.0256	41	.0244	38	.0263	36	.0278	25	.0400

Installation Labor
Neither the duct size nor the percentage of fittings affects the installation productivity rate appreciably. Consequently the erection rate remains relatively constant at 30 Sf/Hr.

Correction Factors

1. Shop Labor
    a. Hand grooving machine versus grooving machine 1.50
    b. Auto grooving machine versus manual machine .... .90
    c. Auto closer versus manual 70
    d. Auto groover and closer 60
    e. Exhaust and return ducts with tie rods in conduits 1.33
    f. Fittings versus straight duct 1.50
    g. "V" groove versus shiplap 1.05

2. Channel or "T" bar reinforcing versus tie-rods
    Shop 1.20
    Field   1.10

3. 475 Board versus 800,   Material costs 90
                               Shop and field labor.....................................95

4. 1400 Board versus 800,   Material costs 1.20
                               Shop and field labor.................................1. 05

5. Material quantity discounts 0-2400 SF..1.05
        2400-20,000 Sf 1.00
        Factory, truck load 94
        Factory, car load 90

Figure 19-12
Rectangular Black Iron Ductwork
With Companion Angles Per Piece Labor and Material

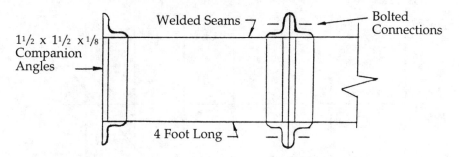

Uses boiler breechings, oven exhausts, material conveying, kitchens

Duct Size Inches	Semi-Perim Inches	Lb/Lf W/20% Waste	Angle Wt	Wt 4 Ft Pc with Angles	Shop Labor Hr/Pc	Field Labor Hr/Pc
12×12	24	14.6	13	71	2.6	1.7
18×12	30	18.2	15	88	3.0	1.9
24×12	36	21.9	18	106	3.4	2.1
30×12	42	25.5	21	123	3.8	2.5
36×12	48	29.3	23	140	4.2	2.8
36×18	54	32.9	26	158	4.7	3.2
42×18	60	36.6	28	174	5.0	3.2
48×18	66	40.2	31	192	5.6	3.5
54×18	72	43.8	34	209	6.2	3.7
60×18	78	47.5	36	226	6.7	4.2
66×18	84	51.1	39	243	7.1	4.7
72×18	90	54.8	41	260	7.4	5.3
72×24	96	58.6	44	278	7.9	5.8
78×24	102	62.4	46	293	8.3	6.3
84×24	108	65.8	50	329	8.6	6.8

Correction Factors (Multipliers on 14 gauge per piece labor)

	Shop	Field
1. 18 gauge	.80	.80
2. 16 gauge	.90	.90
3. 12 gauge	1.10	1.25
4. 10 gauge	1.20	1.45
5. 3/16ths inch plate	1.60	2.60
6. 5 foot long joints	1.10	1.10
7. 8 foot long joints	1.50	1.50

Budget Price, Typical 14 gauge 24×12 duct section, $2.63 per lb.

Figure 19-13
Weight of Aluminum Ductwork per Linear Foot with 20% Allowance

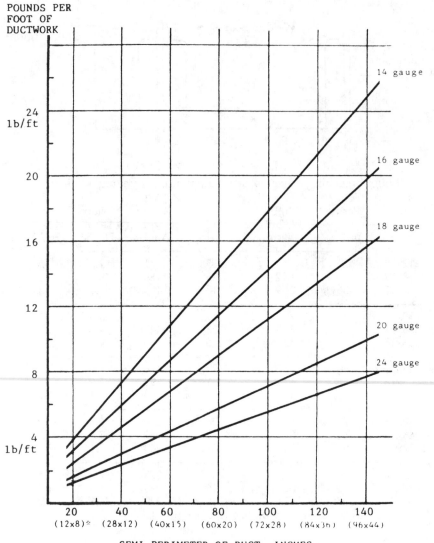

SEMI-PERIMETER OF DUCT, INCHES
(Width plus depth)

* Sample size

Figure 19-14
Budget Estimating External Insulation Per Foot
for Standard HVAC Rectangular Ductwork
25% Fittings by Square Footage

Size	Semi-Perim (w&d) Inches	Sq ft/ft with 15% Waste	Selling Price Furnished & Installed per ft Insulation
6×6	12	2.3	$4.72
12×6	18	3.5	7.18
12×12	24	4.6	9.44
18×9	24	5.2	10.68
18×12	30	5.8	11.90
24×9	33	7.5	15.40
24×12	36	6.9	14.19
24×15	39	7.5	15.40
30×12	42	8.0	16.42
30×18	48	9.2	18.89
30×24	54	10.4	21.33
36×12	48	9.2	18.89
36×18	54	10.4	21.33
36×24	60	11.5	23.61
42×12	54	10.4	21.33
42×18	60	11.5	23.61
42×24	66	12.7	26.07
48×12	60	11.5	23.61
48×18	66	12.7	26.07
48×24	72	13.8	28.32
54×24	78	15.0	30.79
54×30	84	16.1	33.04
54×36	90	17.3	35.51
60×18	78	15.0	30.79
60×24	84	16.1	33.04
60×30	90	17.3	39.81

(Continued)

Figure 19-14 (*Continued*)

Size	Semi-Perim ——— (w&d Inches	Sq ft/ft with 15% Waste	Selling Price Furnished & Installed per ft Insulation
72×24	96	18.4	37.76
72×30	102	19.6	40.23
72×36	108	20.7	42.48
84×30	114	21.9	44.95
84×36	120	23.0	47.21
84×42	126	24.1	49.46
96×24	120	24.2	49.67
96×36	132	25.3	51.93
96×48	144	27.6	56.65
96×72	168	32.2	66.09
96×96	192	36.8	75.51

1. External insulation based on 1-1/2 inch thick 3/4 pound density fiber glass with vapor barrier, at $1.95 per sq. ft. installed.

Figure 19-15
Budget Estimating Insulation and Lining per Foot
for Standard HVAC Rectangular Ductwork
25% Fittings by Square Footage

Duct Size (w × d) Inches	Semi-Perim. (w & d)	Sq Ft/Ft with 15% Waste	Selling Price Furnished & Installed per Ft	
			Insulation	Lining
6 × 6	12	2.3	3.42	2.92
12 × 6	18	3.5	5.20	3.81
12 × 12	24	4.6	6.84	5.83
18 × 9	24	5.2	7.74	6.59
18 × 12	30	5.8	8.62	7.29
24 × 9	33	7.5	11.16	9.52

(*Continued*)

Figure 19-15 (*Continued*)

Duct Size (w × d) Inches	Semi- Perim. (w & d)	Sq Ft/Ft with 15% Waste	Selling Price Furnished & Installed per Ft Insulation	Lining
24 × 12	36	6.9	10.28	8.75
24 × 15	39	7.5	11.16	9.52
30 × 12	42	8.0	11.90	10.15
30 × 18	48	9.2	13.69	11.67
30 × 24	54	10.4	15.46	13.19
36 × 12	48	9.2	13.69	11.67
36 × 18	54	10.4	15.46	13.19
36 × 24	60	11 5	17.11	14.59
42 × 18	60	11.5	17.11	14.59
42 × 24	66	12.7	18.89	16.11
48 × 12	60	11.5	17.11	14.59
48 × 18	66	12.7	18.89	15.23
48 × 24	72	13.8	20.52	17.51
54 × 24	78	15.0	22.31	19.03
54 × 30	84	16.1	23.94	20.43
54 × 36	90	17.3	25.73	21.95
60 × 18	78	15.0	22.31	19.03
60 × 24	84	16.1	23.94	20.43
60 × 30	90	17.3	28.85	21.95
72 × 24	96	18.4	27.36	22.84
72 × 30	102	19.6	29.15	24.87
72 × 36	108	20.7	30.78	26.26
84 × 30	114	21.9	32.57	26.64
84 × 36	120	23.0	34.21	29.18
84 × 42	126	24.1	35.84	30.58
96 × 24	120	24.2	35.99	30.70
96 × 36	132	25.3	37.63	45.95
36 × 48	144	27.6	41.05	35.02

(*Continued*)

Figure 19-15 (*Continued*)

Duct Size (w × d) Inches	Semi-Perim. (w & d)	Sq Ft/Ft with 15% Waste	Selling Price Furnished & Installed per Ft	
			Insulation	Lining
96 × 72	168	32.2	47.89	40.85
96 × 96	192	36.8	54.72	46.70

1. External insulation based on 1-1/2 inch thick 3/4 pound density fiber glass blanket with vapor barrier, at $1.47 per sq. ft. installed.
   Lining is based on 1 inch thick 1-1/2 pound density fiber glass blanket at $1.27 per sq. ft. installed. Installed selling prices based on $27.50 per hour for labor, a 25 percent markup for overhead and a 5 percent markup for profit. Includes all pins, tapes, cement, materials and labor.
2. Galvanized ductwork weight must be increased 12 percent to compensate for increasing the metal duct size one inch for the lining.

Figure 19-16
Estimating Acoustic Lining per Sq Ft

Calculating Material
1. If you have the square footages just add 15% for waste and corner
    overlappings
    Material Cost   1" thick, 1-1/2 Lb density ....33¢ /Sf
                          1/2" thick, 2 Lb density ....26¢ /Sf
                          Cement and Pins, add ....7¢ /Sf
    Increased Duct Size — Increase the metal duct size to cover the lining
    thickness. For example, increase a 20 × 10 duct to 22 × 12 for 1"-thick
    lining. Weight increases about 12 percent for average duct sizes for 1-
    inch thick lining and 6 percent for 1/2-inch thick.
    Labor (Based on 1"-thick, 1-1/2 Lb density, blanket.)

Maximum Width of Duct					
Percentage of	0-12"	13-30"	31-54"	55-84"	85" up
Fittings by	26 ga	24 ga	22 ga	20 ga	18 ga
Square Feet		Square Feet per Hour			
of Total					
Str Duct Only	54	70	84	98	112
10 - 20%	46	58	70	82	94
20 - 30%	41	50 *	62	73	83
30 - 40%	38	45	55	66	74
40 - 50%	35	43	50	58	66
Fittings Only	24	30	35	41	47

*Average

Correction Factors on Labor
1. 1" Thick, 3 Lb 1.15
2. 2 Inch Thick, 1-1/2 Lb 1.15
3. Rigid Board, 1" thick 1.50
    Rigid Board 1-1/2" thick 1.75
    Rigid Board 1-1/2" thick 2.00
4. 1/2" Thick Blanket, 1-1/2 Lb 85

Budget Figures
1. Typical lining:1" thick, 1-1/2 LB, 30% markup $1.27/Sf
    1/2" thick, 1-1/2 Lb, 30% markup 1.06/Sf
2. The square footage of lining and wrap on a job equals approximately
    two-thirds of the duct weight.

Figure 19-17
Estimating Insulation per Sq Ft External Wrapping for Ductwork

Item	Thick-ness	Density Lbs/Cf	Labor Sf/Hr	Mat'l Cost per Sf	Sell Price w/O&P per Sf
Fiber Glass Blanket, RFK					
Vapor Barrier or Vinyl	1"	3/4	50	$.31	$1.28
	1-1/2"	3/4	45	.42	1.50
	2"	3/4	40	.55	1.86
Fiber Glass, Rigid Board					
With RFK Vapor Barrier	1"	3	16	.36	3.21
With 8 oz. Canvas Cover	1"	3			3.83
With 8 oz. Canvas & Painted					
	1"	3			4.59
With RFK Vapor Barrier	2"	3	13	.42	4.01
With 8 oz. Canvas Cover	2"	3			4.78
With 8 oz. Canvas and Painted					
	2"	3		5.73	
Outside Fiber Glass, Rigid Board					
With 2 Layers black mastic, glass mesh					
	1"	4	1	3.44	
	2"	4	1	3.83	
Kitchen Exhausts & Breechings					
Calcium Silicate Block					
Rectangular	2"		8	1.37	6.73
Round	2"		7	1.84	7.82
With 1/2" Cement					
Rectangular	2"		5	2.44	10.99
Round	2"		4	3.21	13.83
Fiber Glass, Rigid Board	2"	4	8	.90	6.23

*Includes 15% waste, pins, staples, tape, installation, supervision, 20% overhead and 5 percent profit. Based on union wages of $37.00 per hour, 10 foot high ducts, lower floors, normal space conditions.

Sf: Sq Ft   Cf: cu ft

Correction Factors
1. Congested ceiling spaces, areas 1.15

DUCTWORK AND ACCESSORIES
Check Off List

Galvanized
— Rectangular
— Spiral
— Residential Round
— Low Pressure
— Medium/High Pressure

Connections
— Cleats
— Angle Flanges
— Ductmate
— TDC Bent Metal Flanges
— Sealed

Insulation
— Lining
— External Insulation

Fiberglass Ductboard
— 475 Board
— 800 Board
— 1400 Board
— Angle or Channel Reinforcing
— Wire Rod Reinforcing

Flexible Tubing
— Bare Low Pressure Insulated L.P.
— Bare High Pressure
— Insulated H.P.
— Metal Flex
— Spin In Collars

Industrial Exhaust Ductwork, Heavy Gauges
— Black Iron
— Stainless Steel
— Aluminum
— Galvanized
— PVC Plastic
— FRP
— Transite
— Rectangular
— Round
— Angle Flange Connections
— Bent Metal Flanges
— Welded Connections
— Strap Hangers
— Rods
— Trapeze Angles
— Joist Spanning Angles
— Gat Clips
— C Clamps

Chapter 20

Estimating Pumps, Piping And Insulation

This chapter covers estimating tables on pumps, bronze and iron valves and balancing devices, as well as budget estimating installed per foot pricing on steel piping and copper tubing.

Heat loss tables for steel pipe, copper tubing and steam piping are included plus a graph for piping heat gains for cooling.

Figure 20-1. Chilled and hot water pumps
Heavy Duty Cast Iron, 3500 RPM

GPM	Approx Head Ft	HP	Direct Material Cost		Labor	Total Material & Labor	
			Each	Per Gal	Man Hours	Direct Cost	With 30% O&P
20	40	1/2	$447	$22.36	2	$521	$677
40	50	1	763	19.08	3	874	1,136
60	60	2	1,056	17.60	4	1,204	1,565
80	70	3	1,140	14.25	5	1,325	1,722
100	80	5	1,223	12.23	6	1,445	1,878
200	80	7-1/2	1,960	9.80	8	2,256	2,932
300	80	10	2,652	8.84	10	3,022	3,929
400	90	15	2,864	7.16	12	3,308	4,300
500	100	20	3,061	6.12	14	3,579	4,652
600	100	25	3,255	5.43	16	3,847	5,002
800	110	30	3,812	4.76	18	4,478	5,821
1,000	120	30	4,370	4.37	20	5,110	6,644
1,200	120	50	4,746	3.95	22	5,560	7,228
1,400	120	50	4,822	3.44	24	5,710	7,423

Correction Factors
1.   1,750 RPM pumps, material   1.50
2.   All bronze body, material   1.50
Direct labor costs are $37.00 per hour.

Figure 20-2
Budget Estimating Copper Tubing
Installed Price Per foot
50/50 Solder, 300F
Includes Average Number of Fittings, Couplings, Hangers and Solder.
Standard Installation Conditions, Lower Floors, 10 ft High

DIA Inches	L	M	K	DWV
1/2	$6.30	$5.67	$6.78	5.95
3/4	7.99	7.19	8.94	7.55
1	10.19	8.96	11.29	9.41
1-1/4	12.82	11.54	15.82	12.13
1-1/2	14.66	13.20	16.06	13.86
2	18.92	17.02	20.75	17.88
2-1/2	25.29	22.76	27.70	23.89
3	30.79	27.72	33.72	29.10
4	43.99	39.59	51.32	41.57
5	79.19	71.27	93.67	74.83
6	104.11	93.70	135.14	98.39
8	190.63	154.41	226.93	161.62

Prices include-all material costs, labor at $39.00 per hour and overhead and profit markup.

Correction Factors
1.   95/5 solder   1.07
2.   Silver solder   1.16

Figure 20-3
Budget Estimating Steel Piping
Installed Price Per Foot
Includes Average Number of Fittings, Couplings, Hangers,
Standard Installation Conditions, Lower Floors, 10 Ft. High

Dia Inches	Screwed Sch. 40	Welded Sch. 40	Flanged Sch. 40	Grooved Sch. 40
1/2	$5.77	$ —	$ _	$ —
3/4	6.08	—	—	5.20
1	7.26	9.43	—	5.72
1-1/4	8.27	10.34	—	6.24
1-1/2	9 36	11.14	—	7.80
2	12 06	12.60	22.93	10.30
2-1/2	16.00	16.63	27.14	13.00
3	19.58	20.21	31.20	15.60
4	25.67	26.84	42.12	20.64
5	40.56	37.32	53.04	27.14
6	49 92	45.23	65.00	36.19
8	67 08	56.35	84.76	54.71
10	78.00	64.74	126.36	73.55
12	113.88	87.79	171.60	91.73
14	—	—	199.68	—
16	—	—	230.88	—
18	—	—	249.60	—
20	—	—	283.92	—
22	—	—	293.28	—
24	—	—	309.92	—

Prices include all material costs, labor at $39.00 per hour and overhead and profit markup.

Figure 20-4
Threaded Bronze and Iron Valves

Size	Type of Material	Direct Matl Costs	Labor Man Hrs	Total Matl & Labor Direct Costs	With 30% O & P
Gate					
1/2	Bronze	15.40	0.65	32.95	42.84
3/4	Bronze	19.95	0.72	39.39	51.21
1	Bronze	24.15	0.78	45.21	58.77
1-1/4	Bronze	32.40	0.91	56.97	74.06
1-1/2	Bronze	41.23	0.98	67.69	88.00
2	Bronze	57.43	1.04	85.51	111.16
2-1/2	Iron	191.43	1.80	240.03	312.04
3	Iron	235.60	1.90	286.90	372.97
4	Iron	294.50	2.00	348.50	453.05
5	Iron	0.00	0.00	0.00	0.00
6	Iron	0.00	0.00	0.00	0.00
8	Iron	0.00	0.00	0.00	0.00
Check					
1/2	Bronze	16.50	0.65	34.05	44.27
3/4	Bronze	18.68	0.72	38.12	49.56
1	Bronze	24.45	0.78	45.51	59.16
1-1/4	Bronze	34.50	0.91	59.07	76.79
1-1/2	Bronze	39.00	0.98	65.46	85.10

Size	Material				
2	Bronze	60.00	1.04	88.08	114.50
2-1/2	Iron	144.00	1.80	192.60	250.38
3	Iron	172.50	1.90	223.80	290.94
4	Iron	262.50	2.00	316.50	411.45
5	Iron	0.00	0.00	0.00	0.00
6	Iron	0.00	0.00	0.00	0.00
8	Iron	0.00	0.00	0.00	0.00
Ball					
1/2	Bronze	5.32	0.65	22.87	29.73
3/4	Bronze	8.48	0.72	27.92	36.30
1	Bronze	10.88	0.78	31.94	41.52
1-1/4	Bronze	18.98	0.91	43.55	56.62
1/1-2	Bronze	24.15	0.98	50.61	65.79
2	Bronze	30.00	1.04	58.08	75.50
2-1/2	Iron	0.00	1.80	48.60	63.18
3	Iron	0.00	1.90	51.30	66.69
4	Iron	0.00	2.00	54.00	70.20
5	Iron	0.00	0.00	0.00	0.00
6	Iron	0.00	0.00	0.00	0.00
8	Iron	0.00	0.00	0.00	0.00
Globe					
1/2	Bronze	21.98	0.65	39.53	51.39
3/4	Bronze	26.93	0.72	46.37	60.28
1	Bronze	45.00	0.78	66.06	85.88
1-1/4	Bronze	64.50	0.91	89.07	115.79
1-1/2	Bronze	79.50	0.98	105.96	137.75

(Continued)

Figure 20-4 (Continued)

Size	Type of Material	Direct Matl Costs	Labor Man Hrs	Total Matl & Labor Direct Costs	With 30% O & P
2	Bronze	121.50	1.04	149.58	194.45
2-1/2	Iron	367.50	1.80	416.10	540.93
3	Iron	412.50	1.90	463.80	602.94
4	Iron	637.50	2.00	691.50	898.95
5	Iron	0.00	0.00	0.00	0.00
6	Iron	0.00	0.00	0.00	0.00
8	Iron	0.00	0.00	0.00	0.00

Angle

Size	Type of Material	Direct Matl Costs	Labor Man Hrs	Total Matl & Labor Direct Costs	With 30% O & P
1/2	Bronze	30.00	0.65	47.55	61.82
3/4	Bronze	39.00	0.72	58.44	75.97
1	Bronze	55.50	0.78	76.56	99.53
1-1/4	Bronze	73.50	0.91	98.07	127.49
1-1/2	Bronze	97.50	0.98	123.96	161.15
2	Bronze	150.00	1.04	178.08	231.50
2-1/2	Iron	0.00	1.80	48.60	63.18
3	Iron	0.00	1.90	51.30	66.69
4	Iron	0.00	2.00	54.00	70.20
5	Iron	0.00	0.00	0.00	0.00
6	Iron	0.00	0.00	0.00	0.00
8	Iron	0.00	0.00	0.00	0.00

Direct labor costs are $37.00 per hour.

Figure 20-5
Balancing Devices

Venturi Meters

DIA. Inches	GPM Range	Direct Costs Each	Labor Man hrs	Total Matl & Labor Direct Costs	Total Matl & Labor With 30% O & P
1/2	0.2 - 4	62.00	0.60	78.20	101.66
3/4	0.5 - 6	65.00	0.66	82.82	107.67
1	2 - 15	70.00	0.72	89.44	116.27
1-1/4	4 - 23	78.00	0.84	100.68	130.88
1-1/2	6 - 30	88.00	0.90	112.30	145.99
2	8 - 50	102.00	0.96	127.92	166.30
2-1/2	12 - 70	132.00	2.10	188.70	245.31
3	20 - 100	174.00	2.30	236.10	306.93
4	30 - 180	218.00	2.40	282.80	367.64
5	70 - 400	281.00	2.60	351.20	456.56
6	150 - 700	326.00	3.00	407.00	529.10
8	200 -1000	493.00	3.60	590.20	767.26
10	300 -1750	1230.00	4.20	1343.40	1746.42

Figure 20-6
Circuit Setters

DIA. Inches	GPM Range	Direct Costs Each	Labor Man hrs	Total Matl Direct Costs	& Labor With 30% O & P
3/4	0.5 - 6	35.00	0.66	52.82	68.67
1	2 - 15	40.00	0.72	59.44	77.27
1-1/2	6 - 30	62.50	0.90	86.80	112.84
2	8 - 50	90.00	0.96	115.92	150.70
2-1/2	12 - 70	170.00	2.10	226.70	294.71
3	20 - 100	270.00	2.30	332.10	431.73
4	30 - 180	375.00	2.40	439.80	571.74

Labor based on new piping installation. Cutting into existing lines may double or triple the labor. Direct labor costs are $37.00 per hour.

Figure 20-7
Heat Losses from Bare Steel Pipe

Horizontal Pipes

Diam. of Pipe, Inches	Temperature or Pipe, Deg. F										
	100	120	150	180	210	240	270	300	330	360	390
	Temperature Difference, Pipe to Air, Deg. F.										
	30	50	80	110	140	170	200	230	260	290	320
	Heat Loss per Lineal Foot of Pipe, Btu per Hour										
1/2	13	22	40	60	82	106	133	162	193	227	265
3/4	15	27	50	74	100	131	163	199	238	280	325
1	19	34	61	90	123	160	199	243	292	343	399
1-1/4	23	42	75	111	152	198	248	302	362	427	496
1-1/2	27	48	85	126	173	224	280	343	410	483	563
2	33	59	104	154	212	275	344	420	503	594	692
2-1/2	39	70	123	184	252	327	410	502	600	709	827
3	46	84	148	221	303	393	493	601	721	852	994
3-1/2	52	95	168	250	342	444	556	680	816	964	1125
4	59	106	187	278	381	496	621	759	911	1076	1257
5	71	129	227	339	464	603	755	924	1109	1311	1532

Figure 20-7 (Continued)
Horizontal Pipes

Diam. of Pipe, Inches	Temperature or Pipe, Deg. F										
	100	120	150	180	210	240	270	300	330	360	390
	Temperature Difference, Pipe to Air, Deg. F.										
	30	50	80	110	140	170	200	230	260	290	320
	Heat Loss per Lineal Foot of Pipe, Btu per Hour										
6	84	151	217	398	546	709	890	1088	1306	1544	1806
8	107	194	141	509	697	906	1137	1391	1671	1977	2312
10	132	238	420	626	857	1114	1399	1714	2060	2417	2852
12	154	279	491	732	1003	1305	1640	2009	2415	2860	3346
14	181	326	575	856	1173	1527	1918	2350	2826	3347	3918
16	203	366	644	960	1314	1711	2149	2634	3168	3753	4395
18	214	385	678	1011	1385	1802	2266	2777	3339	3958	4635
20	236	426	748	1115	1529	1990	2501	3066	3690	4373	5123

Figure 20-7 (Continued)
Vertical Pipes

Diameter Of Pipe Inches	Temperature of Pipe, Deg. F.										
	100	120	150	180	210	240	270	300	330	360	390
	Temperature Difference, Pipe to Air, Deg. F.										
	30	50	80	110	140	170	200	230	260	290	320
	Heat Lose per Lineal Foot of Pipe, Btu per Hour										
1/2	11	20	35	52	71	93	116	142	170	201	235
3/4	14	25	44	65	89	116	145	177	213	252	294
1	17	31	55	81	111	145	181	222	266	315	368
1-1/4	22	39	69	103	141	183	230	300	337	398	465
1-1/2	25	45	79	118	161	210	263	321	386	456	532
2	31	56	99	147	201	262	328	401	481	569	665
2-1/2	37	68	120	178	244	317	397	486	583	687	805
3	46	83	146	217	297	386	484	592	710	839	980
3-1/2	52	94	166	248	339	440	552	676	810	958	1119
4	59	106	187	279	382	496	622	760	912	1078	1259
5	72	131	231	344	472	612	768	939	1126	1331	1555
6	86	156	275	410	562	729	915	1119	1342	1587	1853
8	112	203	358	534	731	950	1191	145	1747	2065	2412
10	140	254	447	667	913	1186	1487	1818	2181	2578	3012
12	166	301	530	790	1081	1404	1761	2154	2584	3054	3567
14	195	354	624	930	1273	1653	2073	2536	3042	35 96	4200
16	221	400	705	1051	1438	1868	2343	2865	3437	4063	4745
18	234	425	748	1115	1526	1982	2486	3040	3648	4311	5036
20	260	472	831	1239	1696	2203	2763	3378	4053	4791	559

Figure 20-8
Heat Losses from Bare Copper Tube

Horizontal Tubes

Nominal Diameter of Tube, Inches	Temperature of Tube, Deg. F.										
	100	120	150	180	210	240	270	300	330	360	390
	Temperature Difference, Tube to Air, Deg. F.										
	30	50	80	110	140	170	200	230	260	290	320
	Heat Loss per Lineal Foot of Tube, Btu per Hr										
1/4	4	8	14	21	29	37	46	56	66	77	88
3/8	6	10	18	28	37	48	60	72	85	99	114
1/2	7	13	22	33	45	59	72	88	104	121	139
5/8	8	15	26	39	53	68	85	102	121	141	163
3/4	9	17	30	45	61	79	97	117	139	162	187
1	11	21	37	55	75	97	120	146	173	201	232
1-1/4	14	25	45	66	90	117	145	175	207	242	279
1-1/2	16	29	52	77	105	135	167	203	241	281	324
2	20	37	66	97	132	171	212	257	305	356	411
2-1/2	24	44	78	117	160	206	255	310	309	429	496
3	28	51	92	136	186	240	297	360	428	501	578
3-1/2	32	59	104	156	212	274	340	412	490	573	662
4	36	66	118	174	238	307	381	462	550	644	744
5	43	80	142	212	288	373	464	561	669	783	905
6	51	93	166	246	336	432	541	656	776	915	1059

Figure 20-8 (Continued)

Horizontal Tubes

Nominal Diameter of Tube, Inches	Temperature of Tube, Deg. F.										
	100	120	150	180	210	240	270	300	330	360	390
	Temperature Difference, Tube to Air, Deg. F.										
	30	50	80	110	140	170	200	230	260	290	320
Heat Loss per Lineal Foot of Tube, Btu per Hr											
8	66	120	215	317	435	562	699	848	1010	1184	1372
10	80	146	260	387	527	681	848	1031	1227	1442	1670
12	94	172	301	447	621	802	999	1214	144	1699	1969

Vertical Tubes

Nominal Diameter of Tube Inches	Temperature of Tube, Deg. F.										
	100	120	150	180	210	240	270	300	330	360	390
	Temperature Difference, Tube to Air, Deg. F.										
	30	50	80	110	140	170	200	230	260	290	320

Figure 20-8 (Continued)

Horizontal Tubes

Nominal Diameter of Tube, Inches	\multicolumn{11}{c}{Temperature of Tube, Deg. F.}

Temperature of Tube, Deg. F.

Temperature Difference, Tube to Air, Deg. F.

Heat Loss per Lineal Foot of Tube, Btu per hr.

Nominal Diameter of Tube, Inches	100	120	150	180	210	240	270	300	330	360	390
	30	50	80	110	140	170	200	230	260	290	320
1/4	3	6	10	15	21	27	34	41	49	57	66
3/8	4	8	14	21	28	36	45	55	65	77	88
1/2	5	10	17	26	35	46	57	69	82	96	111
5/8	6	12	21	31	42	54	66	78	98	114	132
3/4	7	14	24	36	49	64	79	96	114	134	155
1	10	18	31	46	63	82	102	123	147	172	198
1-1/4	12	21	38	57	77	100	125	151	180	210	243
1-1/2	14	25	45	67	91	118	147	178	212	248	287
2	18	33	59	88	120	155	192	233	277	325	375
2-1/2	22	41	73	109	148	191	238	288	343	402	464
3	27	49	87	129	176	227	283	343	408	478	552
3-1/2	31	57	101	150	204	264	328	398	474	554	641
4	35	64	114	171	232	300	374	453	539	631	729
5	43	80	142	212	288	373	464	561	669	783	905
6	52	96	170	253	344	445	554	670	798	934	1080
8	69	127	226	337	458	592	737	892	1063	1244	1438
10	86	158	281	410	570	737	917	1110	1322	1548	1789
12	103	189	336	501	682	881	1097	1328	1582	1851	2140

Figure 20-9
Cooling Heat Gain for Various Pipe Sizes
and insulation Thickness 45°F Water

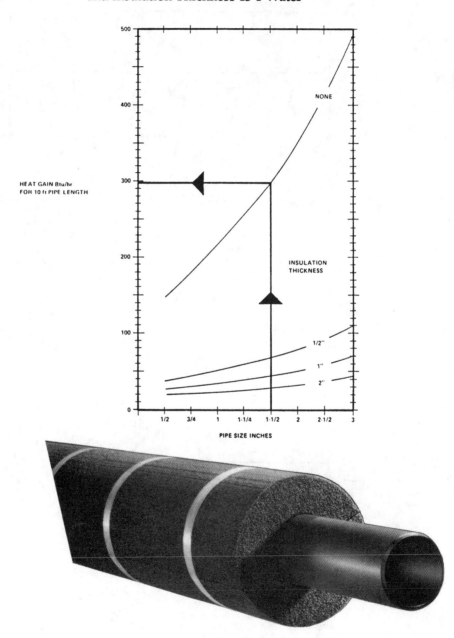

HEAT GAIN Btu/hr
FOR 10 ft PIPE LENGTH

NONE

INSULATION
THICKNESS

1/2"

1"

2"

PIPE SIZE INCHES

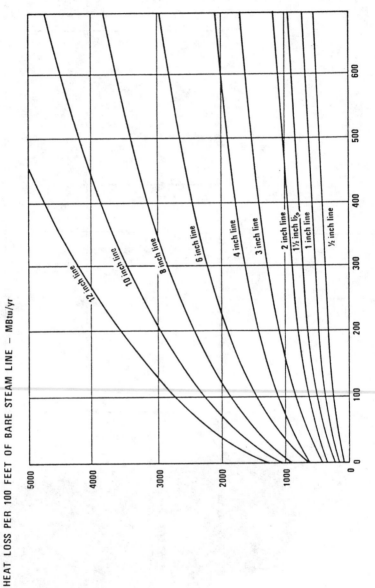

Figure 20-10
Heat Loss From Bare Steam Lines

Figure 20-11. Minimum Insulation Thickness for Piping
ASHRAE 90A-1980 Minimum Insulation Standards—Commercial Piping*

PIPING SYSTEM TYPES	FLUID TEMPERATURE RANGE (°F)	PIPE SIZES					
		Runouts Up to 2"	1" and under	1¼"-2"	2½"-4"	5"-6"	8" and over
		MINIMUM INSULATION (inches) FOR PIPE SIZES					
HEATING SYSTEMS Steam & Hot Water							
High Pressure/Temp	306-450	1.5	2.5	2.5	3.0	3.5	3.5
Medium Pressure/Temp	251-305	1.5	2.0	2.5	2.5	3.0	3.0
Low Pressure/Temp	201-250	1.0	1.5	1.5	2.0	2.0	2.0
Low Temperature	120-200	0.5	1.0	1.0	1.5	1.5	1.5
Steam Condensate for feed water	Any	1.0	1.0	1.5	2.0	2.0	2.0
COOLING SYSTEMS Chilled Water	40-55	0.5	0.5	1.0	1.0	1.0	1.0
Refrigerant, or Brine	Below 40	1.0	1.0	1.5	1.5	1.5	1.5

* These standards are based on insulation
  having thermal resistivity in the range
  of 4.0 to 4.6 ft²•h•deg.F/Btu•in on a
  flat surface at a mean temperature of 75F.

Piping And Valves
Check Off List

Piping
— Steel
— Copper
— Stainless Steel
— PVC
— FRP
— Glass
— Brass
— Screwed Connection
— Flanged
— Welded
— Grove

Hangers
— Split Ring
— Roller
— Spring
— Isolation
— Pipe Saddle
— Riser Clamp
— Strap
— U Bolt
— Adjustable Band

Steam Traps
— Float & Thermostatic
— Bellows
— Thermodynamic
— Inverted Bucket

Flow Measuring Valves
— Venturi
— Circuit Setter
— Pitot Tube
— Illinois Balancing Valve
— Balancing Cock

Meters
— Single Port Meter
— Differential Meter

Special Labor
— Drain System
— **Flush and Clean**
— Pressure Test
— Seal Leaks
— Remove Piping
— Reinstall Piping

Insulation
— Hot Water Supply
— Steam Supply
— Boiler

Control Valves
— 2 Way
— 3 Way Diverting
— 3 Way Mixing

Valves
— Gate   — Plug
— Angle   — Pinch
— Globe   — Relief
— Check   — Safety
— Butterfly   — Strainer
— Ball   — Air Eliminator

Section 12

Controls and Electrical

Chapter 21 -

Temperature Controls and Energy Management Systems

Chapter 22 -

Lighting and Electrical Energy Savings

Chapter 21

Temperature Controls
And
Energy Management Systems

This chapter covers pneumatic control systems and Energy Management Systems(EMS).

There are budget estimating tables for estimating overall pneumatic system costs per horse power and as a percentage of equipment costs, tables for estimating control valves and a temperature control retrofit check off list.

The second section of this chapter covers energy management systems, descriptions of central and distributed systems, budget prices on EMS's per channel, EMS diagram description of EMS function sand EMS check off list.

The end of the chapter also discusses mixed and supply air resets.

Figure 21-1
Temperature Controls and Energy Management Systems

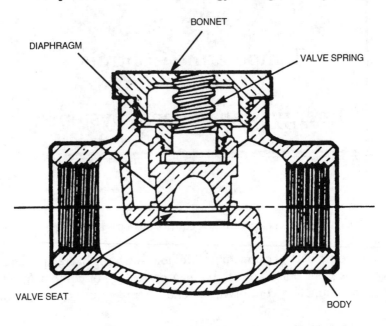

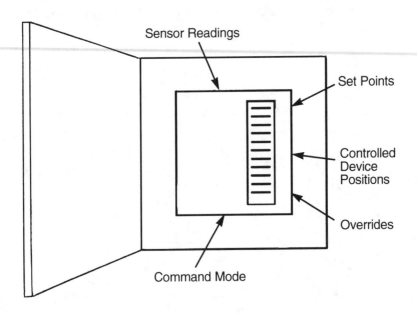

Table 21-1
Budget Control Pricing for Built-up Systems in New Buildings

	Percent of Equipment Costs
Under $25,000	24%
$25,000 to 50.000	20
$60.000 to $100,000	17
$100,000 to $250,000	15
$250,000 to $500,000	12

Add for computerized controls + 10%

Controls for Package Systems

Under $25,000	8%
$25,000 to $100,000	7
$100,000 to $500,000	6

Table 21-2
Pneumatic Temperature Control Compressor Systems Includes Compressor, Tank, Tubing, Dryer, Filter, PRV Valve, Motor, Starter

HP	Direct Material Cost			Labor	Total Material & Labor	
	Duplex Compressor	Other Items	Total Matl Costs	Man Hours	Direct Cost	With 30% O&P
1/2	$1,063	$4,181	$5,244	20	$5,984	$7,779
3/4	1,242	5,520	6,762	21	7,539	9,801
1	1,428	5,679	7,107	23	7,958	10,345
1-1/2	1,750	6,116	7,866	25	8,791	11,428
2	2,022	6,948	8,970	30	10,080	13,104
3	2,342	7,870	10,212	35	11,507	14,959
5	2,625	9,795	12,420	40	13,900	18,070

1. Includes installation of compressor, piping hook up, valves, etc.
2. To furnish and install 3/8" diameter copper trunk line tubing add $6.50 per foot
3. If polyvinyl tubing is used instead of copper tubing deduct 25% from copper tubing.
   Direct labor costs are $37.00 per hour.

Table 21-3
Two Way Control Valves

Dia. Inches	Cv	Direct Costs Each	Type Connect & Press	Labor Man hrs	Total Direct Costs	Matl & Labor With 30% O&P
1/2	.63,1.0,1.6	139.00	Screw, hp	0.55	153.85	200.01
1/2	2.5,4.0	86.00	Screw	0.55	100.85	131.11
3/4	6.3	145.00	Screw,hp	0.60	161.20	209.56
3/4	6.3	89.00	Screw	0.60	105.20	136.76
1	10.0	170.00	Screw,hp	0.65	187.55	243.82
1	10.0	95.00	Screw	0.65	112.55	146.32
1-1/4	16.0	194.00	Screw,hp	0.77	214.79	279.23
1-1/4	16.0	117.00	Screw	0.77	137.79	179.13
1-1/2	25.0	230.00	Screw,hp	0.83	252.41	328.13
1-1/2	25.0	156.00	Screw	0.83	178.41	231.93
2	40.0	212.00	Screw,hp	0.88	235.76	306.49
2	40.0	201.00	Screw	0.88	224.76	292.19
2-1/2	63.0	278.00	Screw	2.00	332.00	431.60
2-1/2	63.0	561.00	Flange	2.00	615.00	799.50
3	100.0	375.00	Screw	2.10	431.70	561.21

Table 21-3 (*Continued*)
Two Way Control Valves

Dia. Inches	Cv	Direct Costs Each	Type Connect & Press	Labor Man hrs	Total Direct Costs	Matl & Labor With 30% O&P
3	100.0	672.00	Flange	2.10	728.70	947.31
4	160.0	930.00	Flange	2.40	994.80	1293.24

Three Way Control Valves

Mixing

Dia. Inches	Cv	Directs Costs Each	Type Connect	Labor Man hrs	Total Matl & Labor Direct Costs	with 30% O&P
1/2	4.0	112.00	Screw	0.55	126.85	164 91
3/4	6.3	124.00	Screw	0.60	140.20	182 26
1	10.0	144.00	Screw	0.65	161.55	210.02
1-1/4	16.0	169.00	Screw	0.77	189.79	246 73
1-1/2	25.0	200.00	Screw	0.83	222.41	289 13
2	40.0	235.00	Screw	0.88	258.76	336.39
2-1/2	63.0	638.00	Flange	2.00	692.00	899.60

Table 21-3 (Continued)

Dia. Inches	Cv	Direct Costs Each	Type Connect & Press	Labor Man hrs	Total Direct Costs	Matl & Labor With 30% O&P
3	100.0	755.00	Flange	2.10	811.70	1055.21
4	160.0	1263.00	Flange	2.40	1327.80	1726.14

Diverting

Dia. Inches	Cv	Directs Costs Each	Type Connect	Labor Man hrs	Total Direct Costs	Matl & Labor with 30% O&P
2-1/2	63.0	795.00	Flange	2.00	849.00	1103.70
3	100.0	891.00	Flange	2.10	947.70	1232.01

Not Included:
1. Control motor
2. Linkages
3. Plug in balance relays
hp: High pressure
Direct labor costs are $37.00 per hour.

Figure 21-2
Energy Measurement
- Steam
- Hot Water
- Chill Water
- Compressed Air
- Natural Gas

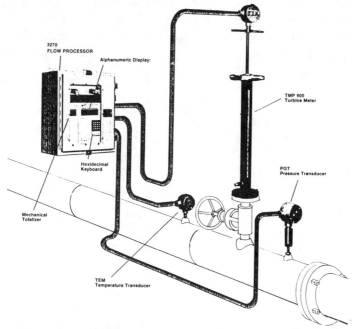

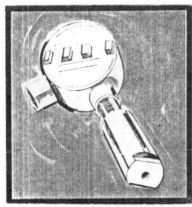

**Pressure
Transducer**

**Temperature
Transmitter**

TEMPERATURE CONTROLS
Check Off List

Thermostats & Temperature Sensors
— Space Thermostat
— Set back Thermostat
— Discharge Duct
— Mixed Air
— Outside Air
— Return Air
— Freezstat
—Remote Bulb Stat

Aquastats —
— Chiller Water Discharge
— Chiller Water Return
— Cooling Tower In
— Cooling Tower Out

Thermometers
— Stem —
— Deep Well

Humidistats

Pressure Sensors
— Pitot Tube

Valves
— 2 Way Control
— 3 Way Diverting
— 3 Way Mixing
— Pressure Reducing Valve
— Radiator
— Solenoid
— Gas Valve
— Gas Valve, Combination
— Thermostatic Expansion Valve

Dampers
— Opposed Black

Panels
— Control
— Readout

Pneumatic Systems
— Copper Tubing
— Plastic Tubing
— Pressure Control Valves
— Filter
— Motor
Starter

Electronic Systems
— Relays
— Wiring
— Capacitors
— Resistors
Timers
— Electronic Ignition
— Intermittent Ignition
— Thermocouples
— Sequences
— Transformers
— Contactors
— Pressure Controls

Control Motors
— 2 Position
— Modulating

Economizers
— Outside Air/Mixed Air
— Enthalpy Control
— Boiler

Safety Switches
— High Temperature Limit
— Low Temperature Limit

— Parallel Blade
— Inlet Vane, Fans
— Combustion Air

VAV Controls
— Transducers
— Transmitters
— Static Pressure Controller
— Control Panel

— High Pressure Limit
— Low Pressure Limit
— Fan and Limit Control, Furnace
— Air Flow Switch
— Sail Switch

ENERGY MANAGEMENT SYSTEMS

Centralized Energy Management Systems

A centralized system is one in which all processing and data storage takes place in a single computer unit. This computer interfaces to sensors, local loop controllers and display devices through multiplexing or remote field processing units. The remote field processing units are slaved to a host central processing unit and have no stand-alone capability. If a remote processor is capable of performing its local functions independent of other processors, it is considered a stand-alone.

The central processing unit maintains a fully replicated database, while each field processing unit typically maintains a local point database. Changes in field point conditions are directed to the host central processing unit via change-of-state messages. The field processing units can execute field commands, such as resetting the setpoint of a local loop controller, when directed to do so by the host central processing unit.

A centralized system, by having all function software and data at one location, allows for rapid and integrated action. Thus, for instance, demand limiting can decide if it has a higher priority than the interlocking feature by simply examining the common point database. A disadvantage of this structure is that all capability ceases upon failure of the central processing unit unless a redundant processor is included.

Distributed Energy Management System, Type I

In a distributed processing system with centralized data, the central processing unit performs global system functions such as demand limiting, global interlocks, log and report generation. Stand-alone processors provide localized monitoring and control functions which do not require centralized coordination such as duty cycling, optimal start/stop, local interlocking, time programming, direct digital control. User access to local data or program software at the remote processor generally is provided through a keypad or portable terminal.

The point capacity and functionality of the stand-alone processor often is matched to functional or area subsystems within a facility, such as an air handling unit or a mechanical equipment room.

The stand-alone processor maintains a local database that contains interlock tables, time programmed, etc. The host central processing units maintains a fully replicated, centralized database. Using a replicated database allows for lower-speed communications technologies such as twisted-pair and digital PBXs (private branch exchanges for telephone networks).

Stand-alone processors may communicate with the central processing unit, but not with each other. However, the processors may request global information such as outdoor temperature from the central processing unit.

Distribution system functions help avoid communication peak problems whereas centralized systems can become bogged down with data during some circumstances.

Equipment duty cycling takes place in the remote processor after receiving a duty cycle schedule from the central processing unit.

Energy management strategies such as optimal start/stop and supply air reset are local in nature, but may need data that are not local to the particular processor.

Distributed Energy Management System, Type II

The second type of distributed energy management system with distributed data consists of an optional host central processing unit interconnected to stand-alone remote processors via a common communications bus where any node can "talk" to any other node. This scheme also supports a general broadcast capability, which is the ability of a processor to communicate simultaneously with all other nodes on the network.

The central processing unit maintains a directory of the field points, however, complete field point data not stored such as global data and database download information.

Implementation of the optimal start/stop feature is used as an example to describe communication between remote processors in the distributed network.

In the distributed database architecture, the remote processor needing data "broadcasts" a request for the information. The remote node that monitors the needed field point data intercepts the request and formats a reply message with the current point data.

DISTRIBUTED STAND ALONE UNIT SYSTEMS

Stand alone unit systems run 10 to 20 percent more than centrally-controlled systems.

However, the extra costs, are offset by the following advantages:

Easier programing.

Less labor costs in maintenance, repair and troubleshooting.

Easier to add on system without extensive wiring.

Control is lost in centrally-controlled or partially distributed systems if the central computer fails.

ENERGY MANAGEMENT SYSTEMS DEFINITIONS

1. Analog Inputs
    Monitors variables such as temperature, humidity
    and pressure.

2. Analog Outputs
    Drives, valves, dampers and motors.

3. Digital Inputs
    On/off type readings for: flow/no *flow* on/off alarms high/low limit
    alarms

4. Digital Outputs
    Turns equipment on/off.
    Overrides for air handling units, fans etc.

Energy Management Systems
(Indirect, Low Voltage, Hardwire)

Equipment Prices
>   10 Channel EMS unit W7010H* $4,167
>   20 Channel EMS unit W7020H* $6,834

Installed Prices
>   10 Channel @ $1,200/pt $16,080
>   20 Channel @ $1,000/pt $26,800

Communications Link Hardware $1,608
Remote Computer
>   For Programing and Monitoring $2,680 to $4,020
>   (Can put whole program on a floppy disk)

Functions
- Time of Day Scheduling
- Optimum Start/Stop
- Demand Limiting
- Duty Cycling, Temperature Compensated by Temperatures in spaces
- Monitoring Outside Air and Indoor Air Temperatures

*Approximate Honeywell Contractor Price
Direct labor costs are $37.00 per hour.

Figure 21-3
Distributed Control

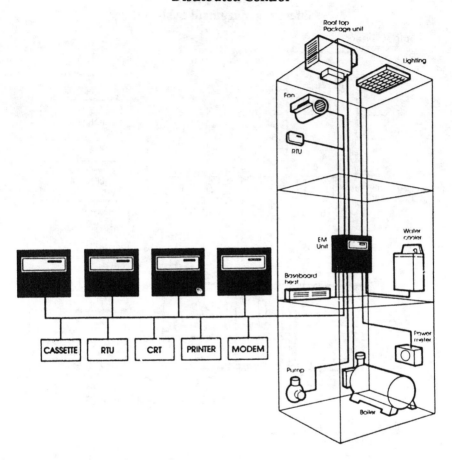

Figure 21-4
Direct Digital
Energy management System

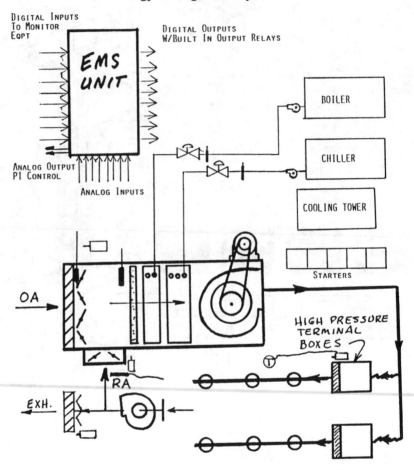

Figure 21-5
D.D.C. Components

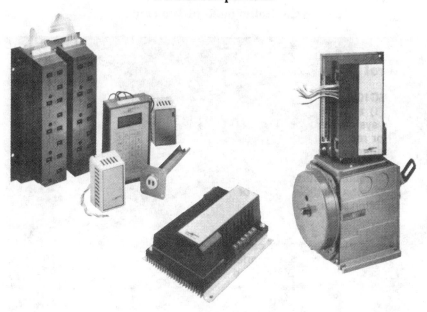

Figure 21-6
Energy Management System Functions

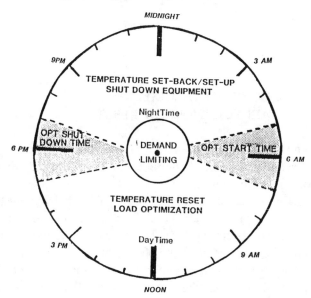

FACTORY INSTALLED E.M.S. CONTROLLER ON ROOF TOP UNITS

Figure 21-7
Factory Installed E.M.S.
Controller on Roof Top Units

Function: Payback 2 yrs
Scheduling
Demand Limiting
Duty Cycling
Reporting

AUTOMATIC CONTROL OF MIXED
AND SUPPLY AIR TEMPERATURE RESETS

An energy management system can automatically control mixed air temperature and discharge air temperature resets for reducing recooling and reheating energy waste.

Normally, mixed air temperature is controlled during the heating season by mixing outdoor air, at 60 F or lower, and recirculated air to maintain a constant temperature. The mixed air temperature generally ranges somewhere between 50 and 60 degrees.

The purpose in having a mixed air temperature this low in the heating season is to meet the cooling needs of those spaces requiring cooling the year around. Such spaces might be ones which have relatively high sun loads or interior zones with lighting, people and equipment loads.

When the mixed air is kept at a constant temperature, it must be kept low enough to meet the maximum cooling load ever expected in the space requiring the greatest amount of cooling. This temperature is maintained regardless of whether the spaces served actually need it or not. At times when the spaces do not need cooling the system is wasting energy because excess heating energy must be used to overcome the excess cooling effect being introduced by the outside air.

Minimizing Mixed Air Energy Waste

Thus, the mixed air temperature should optimally be reset to match the actual loads in the various spaces. This is done by sensing the spaces that require the greatest amount of cooling, at any given time, and resetting the mixed air temperature upwards until the needs of that space or zone are just being met.

With multizone systems, sensing the zone requirements is easily done because the position of the hot and cold discharge deck dampers are an indication of supply temperature required.

Minimizing Cold Supply Air Energy Waste

The concept behind cold deck discharge air temperature reset is similar to that of the mixed air temperature reset. Match the amount of cooling supplied to the amount actually required by, resetting the cold deck discharge temperature upward whenever possible.

AUTOMATIC CONTROL OF MIXED AND
SUPPLY AIR TEMPERATURE RESETS

An energy management system can automatically control mixed air temperature and discharge air temperature resets for reducing recooling and reheating energy waste.

Normally, mixed air temperature is controlled during the heating season by mixing outdoor air, at 60° F or lower, and recirculated air to maintain a constant temperature. The mixed air temperature generally ranges somewhere between 50 and 60 degrees.

The purpose in having a mixed air temperature this low in the heating season is to meet the cooling needs of those spaces requiring cooling the year around. Such spaces might be ones which have relatively high sun loads or interior zones with lighting, people and equipment loads.

When the mixed air is kept at a constant temperature, it must be kept low enough to meet the maximum cooling load ever expected in the space requiring the greatest amount of cooling. This temperature is maintained regardless of whether the spaces served actually need it or not. At times when the spaces do not need cooling the system is wasting energy because excess heating energy must be used to overcome the excess cooling effect being introduced by the outside air.

Minimizing Mixed Air Energy Waste

Thus, the mixed air temperature should optimally be reset to match the actual loads in the various spaces. This is done by sensing the spaces that require the greatest amount of cooling, at any given time, and resetting the mixed air temperature upwards until the needs of that space or zone are just being met.

With multizone systems, sensing the zone requirements is easily done because the position of the hot and cold discharge deck dampers are an indication of supply temperature required.

Minimizing Cold Supply Air Energy Waste

The concept behind cold deck discharge air temperature reset is similar to that of the mixed air temperature reset. Match the amount of cooling supplied to the amount actually required by, resetting the cold deck discharge temperature upward whenever possible.

Chapter 22

Lighting and Electrical Savings

METHODS OF SAVING ENERGY
LIGHTING AND ELECTRICAL ENERGY

This chapter covers:
Methods of Saving Lighting and Electrical Energy
Motor Retrofits
Variable Frequency Motor Controllers, Description, Costs
kWh and Energy Costs of Motors Per Year
Costs Soft Motor Starter
How to Reduce Demand Charges
Reducing Power Factors
Budget Estimating Motors, Starters and Wiring
Motor Data: Amps, Efficiencies, Power Factors, Starters
Lighting and Electrical Check Off List

Reduce Lighting Levels
• Reduce lighting wattage per sq ft of building commensurate with actual current needs.

• Remove selected florescent bulbs and disconnect or cut wires or remove ballasts.

• Completely disconnect selected fixtures and remove.

Electric lighting generates 60 percent more heat per of illumination than sunlight thus requiring more cooling expenditures. This should be evaluated when considering less lighting and more natural light against increased conduction and solar heat gains through windows.

Replace Lighting
• Replace less efficient incandescent lighting with more efficient florescent fixtures.

- Install high efficiency lighting such as high pressure sodium.

- Install higher efficiency florescent bulbs.

- Install higher efficiency ballasts.

- Install electronic ballasts.

Operation and Maintenance
- Turn lights off when spaces are not occupied.

- Clean bulbs, reflectors and lenses yearly.

- When natural light is available in perimeter areas or through skylights etc., turn off lights.

Efficient Lighting System Design
- Use specific task lighting.

- In high ceiling areas, lower lighting fixtures.

- Lower ceilings.

- Split up lighting controls so one control does not operate all lights in larger areas for local control of occupied and unoccupied spaces.

- Install mirror surface reflectors inserts in florescent fixtures for high efficiency lighting. Can save up to 50 percent on electrical costs and lamp replacements, plus up to 25 percent of air conditioning costs.

Outside Lighting
- Reduce outside lighting commensurate with safety.

- Use photo cell control on outdoor lights.

Motors
- Install smaller motor commensurate with actual load so that existing motor doesn't run on partial load causing less efficiency and lower power factor.

- Install more energy efficient motors.

- Install variable speed variable frequency invertor motors.

Starters
- Install solid state soft motor starters.

Figure 22-1
High Efficiency Motors Savings

Increase cross sectional area of stator and rotor conductors to reduce resistance

Decrease the air gap to reduce the magnetizing current

Lengthen rotor and stator cores to reduce magnetic density

Design rotor and stator slots for reduced leakage reactance

Use thinner core laminations or special low-loss steel to reduce core losses

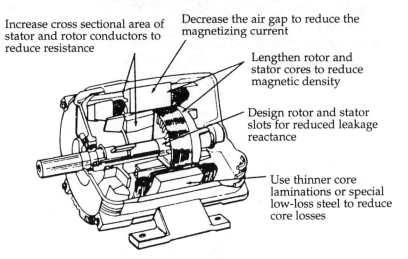

Example:
Given Situation

40 hp motor operating 100 hrs per wk
Existing Motor eff: 88%
High efficiency motor eff: 92%
Existing motor uses 30 kW pwr input
Utility cost: .10/kWh

$$\frac{30 \text{ kW input} \times .88 \text{ eff exist}}{.92 \text{ eff new}} = 28.7 \text{ kW new input}$$

Savings = (30 – 28.7 kW) × (100 hr/wk) × (52 wk/yr) × .10
        = $676/yr

EVALUATING COST-EFFECTIVENESS
OF GROUP MOTOR RETROFITS

1. Consider a group motor retrofit over individually replacing motors because of the shorter composite payback afforded by the more comprehensive retrofit. Lower installation costs and price breaks on bulk equipment orders can shorten the payback period on group retrofits.

2. A group motor retrofit is generally only feasible when the motors are operated a minimum of 3,000 hours per year. Otherwise, the move to energy-efficient motors does not yield enough savings. Ideally, the motors to be changed out should be run 24 hours a day, year round.

3. Evaluate electric rates. Users with rates as low as 2.5 cents per kilowatt hour can realize an attractive return on their investment, provided the motors to be replaced operate a great number of hours.

4. A group retrofit involving standard motors that currently operate at 50 to 70 percent of their load capacity will provide the greatest savings if they are replaced with energy-efficient motors operating at the same load. This is because standard-efficiency motors typically operate at about 65 percent efficiency when they are half-loaded generally while an energy efficient motor operates at closer to 90 percent efficiency at half load.

5. The greatest savings results when motors in the 1 to 50 horsepower (hp) range are changed out. Larger motors from 50 to 75 hp and up yield a smaller improvement in energy efficiency, but a greater number of kilowatt hour savings. Conversely, small motor changeouts involving less than 1 hp result in a greater efficiency change, yet yield fewer kilowatt hour savings because their electricity draw is so insignificant. However, the middle range motors offer an attractive medium between energy-efficiency gains and electricity consumption.

(Per David Lee, president of Lee and Browne Consulting Engineers Inc., Tulsa, and Nancy Grossman, energy management representative at Pacific Gas & Electric Co., San Francisco.)

Figure 22-2. Cost and kW Hours of Motors Based on Full Load AMPS

HP	Full Load AMPS	Voltage	kWh Full Load	kWh Per Year			Cost Per Year @.08 Per kWh		
				Hours Per Year			Hours Per Year		
				8,760	4,380	1,000	8,760	4,380	1,000
1	3.6	230	1.1	9,599	4,800	1,096	$768	$384	$88
2	6.8	230	2.3	20,218	10,109	2,308	1,617	809	185
3	9.6	230	3.2	27,639	13,820	3,155	2,211	1,106	252
5	15.2	230	5.1	44,610	22,305	5,092	3,569	1,784	407
7.5	22	230	7.5	65,564	32,782	7,484	5,245	2,623	599
10	28	230	9.9	86,666	43,333	9,893	6,933	3,467	791
15	42	230	14.5	127,364	63,682	14,539	10,189	5,095	1,163
20	54	230	18.7	164,130	82,065	18,736	13,130	6,565	1,499
25	34	460	23.5	205,734	102,867	23,486	16,459	8,229	1,879
30	40	460	27.8	243,156	121,578	27,758	19,452	9,726	2,221
40	52	460	36.5	319,727	159,864	36,499	25,578	12,789	2,920
50	65	460	46.1	404,191	202,095	46,140	32,335	16,168	3,691
60	77	460	54.8	480,421	240,210	54,843	38,434	19,217	4,387
75	96	460	68.4	598,966	299,483	68,375	47,917	23,959	5,470
100	124	460	89.1	780,580	390,290	89,107	62,446	31,223	7,129
125	147	460	105.9	927,415	463,707	105,869	74,193	37,097	8,470
150	174	460	125.3	1,097,756	548,878	125,315	87,820	43,910	10,025
200	226	460	162.8	1,425,821	712,911	162,765	114,066	57,033	13,021
250	284	460	204.5	1,791,740	895,870	204,537	143,339	71,670	16,363
300	339	460	244.4	2,141,095	1,070,547	244,417	171,288	85,644	19,553

Figure 22-3
Adjustable Frequency Motor Speed Controllers
Variable Frequency Inverter For Fans, Pumps, Chillers Etc.

HP	Weight	Each	Direct Material Per HP	Costs Access.	Labor Man Hrs	Total Matl Direct Costs	& Labor With 30% O&P
1	27	900.00	900.00	725.00	8.60	1857.20	2414.36
1.5	35	1000.00	666.67	725.00	8.60	1957.20	2544.36
2	50	1170.00	585.00	725.00	8.60	2127.20	2765.36
3	72	1340.00	446.67	725.00	8.70	2299.90	2989.87
5	75	1680.00	336.00	731.00	8.80	2648.60	3443.18
7.5	280	2150.00	286.67	737.00	11.00	3184.00	4139.20
10	280	2700.00	270.00	887.00	11.30	3892.10	5059.73
15	600	3500.00	233.33	904.00	11.50	4714.50	6128.85
20	600	4000.00	200.00	931.00	13.50	5295.50	6884.15
25	800	4700.00	188.00	947.00	13.50	6011.50	7814.95
30	800	5600.00	186.67	947.00	14.20	6930.40	9009.52
40	1100	7400.00	185.00	997.00	19.80	8931.60	11611.08
50	1100	9200.00	184.00	1022.00	20.20	10767.40	13997.62
60	1100	11000.00	183.33	1054.00	20.20	12599.40	16379.22
75	2000	14000.00	186.67	1097.00	24.80	15766.60	20496.58

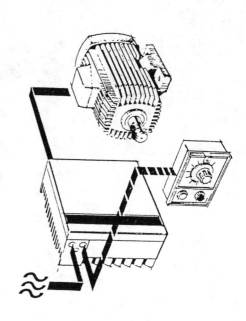

100	2000	25000.00	250.00	1283.00	29.00	27066.00	35185.80
125	2000	28000.00	224.00	1375.00	29.50	30171.50	39222.95
150	2000	32000.00	213.33	1446.00	30.00	34256.00	44532.80
200	2195	37000.00	185.00	1587.00	31.00	39424.00	51251.20
250	2320	40000.00	160.00	1729.00	32.00	42593.00	55370.90
300	2500	44000.00	146.67	2020.00	39.00	47073.00	61194.90
400	2700	48000.00	120.00	2303.00	40.00	51383.00	66797.90
500	2900	50000.00	100.00	2588.00	41.00	53695.00	69803.50

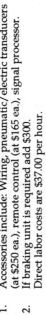

1. Accessories include: Wiring, pneumatic/electric transducers (at $250 ea.), remote control (at $165 ea.), signal processor.
2. If braking unit is required add $300.
    Direct labor costs are $37.00 per hour.

Figure 22-4
Solid State Soft Motor Starters
230 Volt/460 Volt, 3 Phase

HP	Size Starter	Direct Material Cost		Labor	Total Material & Labor	
		Each	Per HP	Man Hours	Direct Cost	With 30% O&P
5	0	$730	$146.00	2.8	$806	$1,047
10	1	800	$80.00	3.1	884	1,149
15	2	890	$59.33	3.5	985	1,280
20	2	960	$48.00	3.5	1,125	1,462
25	2	1,030	$41.20	3.5	1,125	1,462
30	3	1,100	$36.67	4.5	1,222	1,588
40	3	1,300	$32.50	4.5	1,422	1,848
50	4	1,500	$30.00	5.3	1,643	2,136
60	4	1,625	$27.08	5.3	1,768	2,299
75	4	2,250	$30.00	5.3	2,393	3,111
100	5	2,750	$27.50	6.5	2,926	3,803
125		3,250	$26.00		3,250	4,225
150		3,750	$25.00		3,750	4,875
200		4,875	$24.38		4,875	6,338
250		6,000	$24.00		6,000	7,800

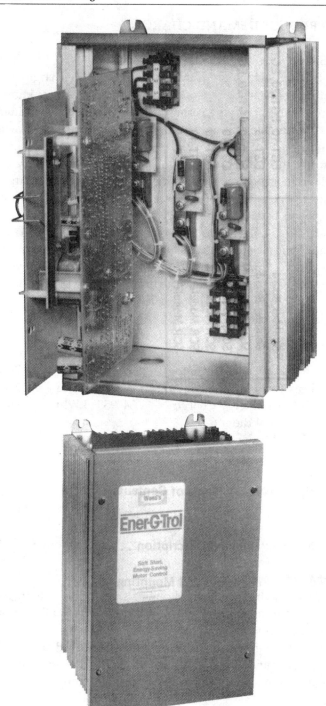

HOW TO REDUCE DEMAND CHARGES

How To Start

The first step in any program to minimize utility charges is to know how the utility rate structure works. In many instances just knowing this and appropriately modifying certain equipment operational time frames can save considerable amounts without any reduction in Kilowatt-hour consumption.

Basically there are two types of charges in most utility company commercial rate structures: The energy charge and the demand charge. The energy charge is based on the direct measurement of kilowatt-hours (kWh) of power used multiplied by the appropriate rate per kWh. The demand charge is a charge to a commercial facility that is based on its highest rate of electric power consumption during a specific billing period. It requires a separate meter from the traditional kWh meter most of us are familiar with, and the rate structure can appear quite controversial if not
properly understood.

Monthly Customer Charge		$484.60
Demand Charge	9532.8 kW	79408.22
Peak Energy Charge	2120049 kWhrs	121054.80
Off-Peak Energy Charge	3409551 kWhrs	$98876.98
Facilities Rental Rider 6		119.22
Meter Rental Rider 7		14.05
State Tax		15237.86
Municipal Tax		16232.58
	Total Bill	$331,428.31

Reflects Fuel Addition of .200 cents per Kilowatthour

How Demand Charge is Measured

The demand charge is the customer's share of the cost of generation capacity and equipment that a utility must have available to serve that customers largest demand for electricity.

To measure demand, the electric meter has two dials; one for kilo-watt-hours (kWh) and one for demand (kW). The demand dials measure the average amount of electricity used every 30 minutes of the monthly billing period.

The meter records the highest average demand, which is read each month before the utility resets the demand dial back to zero.

The demand is based on the average, not the instantaneous, power used during a 30-minute period.

Demand Reduction Methods

Some no cost, low cost manual techniques for demand limiting are:

• Reduce lighting on an overall permanent basis.

• Pull fuses on a couple stages of a multistage electric water heaters particularly in an office building.

• Pull fuses on some electric duct heaters and entrance foyer heaters.

• Schedule high-kW equipment testing on off hours.

Other Demand Reduction Methods

Install energy management system with demand limiting which will turn off or reduce incrementally certain electrical loads for 15 or 30 minutes according to priorities as follows:

Toilet and general exhaust fans
Lighting
HVAC Units
Turn chiller off or limit load
Cut out one or two stages of DX condensors
Cut out one or two stages of electric heating coils

Install variable transformer to reduce the operating voltage and lighting levels of entire lighting areas a direct digital lighting controller is required.

Fit fluorescent fixtures with solid state electronic ballasts that have reset capabilities. The electronic ballast reduces energy and waste heat and can be used.

Install cogeneration system.

Install ice making equipment or chilled water tanks for storage at night. This provides savings, in not only demand charges, but also in using lower off peak electrical rates at night.

Install more energy efficient motors.

Install more energy efficient refrigeration equipment.

Replace electric heating coils with hot water or steam coils and electric boilers with gas or oil fired boilers.

The single most cost-sensitive piece of equipment is the chiller and most have demand limiting on them or can have it applied. Chillers can be locked in at 60%, 70%, or 80%.

Finding Causes of Peaks

Record hourly electrical readings of maximum, average and minimum consumption and plot to determine when peaks occur such as during an early morning startup, peak period of cooling, or other periodic heavy electrical operation.

Some electric companies make available to its commercial customers computerized printouts monthly containing daily/hourly use of kilowatts and percentage of peak in order that administrators and engineers may track down high peaks to cut demand charges as shown on opposite page:

REDUCING POWER FACTOR LOSSES
ON UNDERLOADED MOTORS

Underloaded motors frequently cause poor power factor conditions.

A motor at 90 percent of full load name plate rating may run on the utility companies power factor typically about .85 or .86.

However, if a motor is running at 30 or 40 percent of its full load, it may have a power factor of .70 or .75, depending on its horsepower, which increases the motor electrical operating costs about 30 percent.

Here's an example of the extra costs involved.

A 25 HP HVAC unit motor running full time, 7860 hr per year, at 8 cents a kWh costs about $760 per HP per year for electricity it at full load.

The same motor running at 40 percent of full load at 10 BHP, costs an additional $173 per HP due to the reduced power factor.

This results in an additional $1730 per year for every 100 HP of motors running at the same low loads.

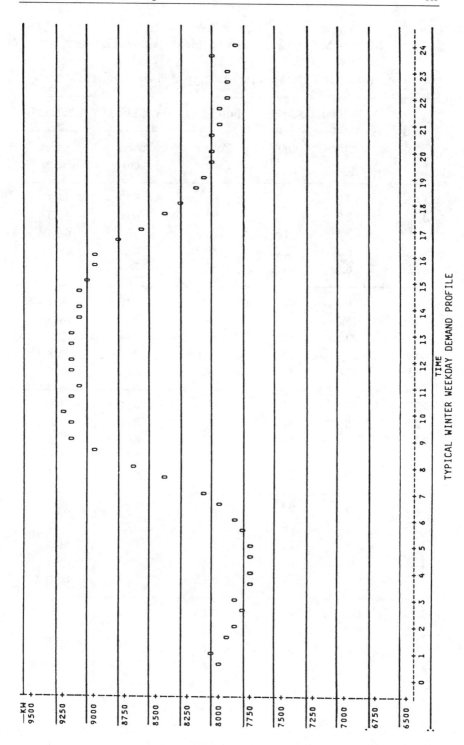

TYPICAL WINTER WEEKDAY DEMAND PROFILE

Figure 22-6
Motor AMP Draws, Power Factor at Different Loads

Induction Type Motors, 1800 rpm, 3 Phase, 60 Cycle

HP	Full Load AMPS		Full Load	3/4 Load	1/2 Load
	230V	460V	Power Factor	Power Factor	Power Factor
1/2	2.0	1.0	69.2	56.2	44.8
3/4	2.8	1.4	72.0	60.0	49.0
1	3.6	1.8	76.5	67.2	54.3
1.5	5.2	2.6	80.5	72.2	59.2
2	6.8	3.4	85.3	76.7	64.3
3	9.6	4.8	82.6	79.5	64.3
5	15.2	7.6	84.2	79.3	69.5
7.5	22	11	85.5	80.4	72.7
10	28	14	88.8	80.3	75.8
15	42	21	37.0	81.4	77.8
20	54	27	87.2	83.2	78.3
25	68	34	86.8	82.8	75.8
30	80	40	87.2	83.5	77.3
40	104	52	88.2	85.7	80.0
50	130	65	89.2	85.7	80.2
60	154	77	89.5	86.3	81.8
75	192	96	89.5	87.2	83.2
100	248	124	90.3	89.5	86.3
125	293	147	90.5	89.5	86.3
150	348	174	90.5	88.8	85.0
200	452	226	90.5		
250	568	284	90.5		
300	678	339	90.6		

Figure 22-7
Budget Estimating Electrical Work

Motors
Induction motors, 1800 Rpm, 230/460 V, 3 phase, 60 cycle, T frame

	Direct Costs of Motor			Labor to Install Motor Hours	Wiring Hookup	
Hp	Dripproof	Totally Enclosed Fan Cooled	Explosion Proof		Labor Hours	Direct Mat. Costs
1/4	$54	$61	$	1.4	.5	$62
1/2	92	98		1.6	.5	62
1	148	173	366	2.0	.6	62
1-1/2	164	191	385	2.2	.6	62
2	177	210	399	2.5	.6	62
3	187	233	459	2.8	.7	62
5	210	271	562	3.0	.8	69
7-1/2	320	386	768	4.0	1.0	75
10	386	466	906	5.0	1.3	75
15	518	623		6.0	1.5	93
20	648	783		7.0	1.5	121
25	768	950		8.0	1.5	137
30	900	1,120		9.5	2.2	137
40	1,130	1,493		12.0	3.8	189
50	1,369	1,898		15.0	4.2	215
60	1,733			18.0	4.2	249
75	2,099			21.0	4.8	293
100	2,698			24.0	5.0	331

Correction Factors
1. Costs of "U" frame motors: dripproof and TEFC 1.3
     explosion proof 1.15

(Continued)

Starters
1. Combination magnetic, fusible disconnect, HOA, pushbutton, transformer, pilot light, MEMA 1 enclosure.

Size	Direct Cost of Starter	Labor to Install & Hookup Hrs	Totally Installed Price W/Hookup O&P
00	$203	2.5	$385
0	267	2.8	471
1	330	3.1	625
2	396	3.5	737
3	692	4.5	1,187
4	1,285	5.3	1,954

2. Manual fractional horsepower starter with overload heaters $13.65
3. Manual starter 1 to 10 hp, overload, start-stop buttons, HOA $57.75

METHODS OF HANDLING EXCESSIVE POWER FACTOR LOSSES AND CAPACITORS

1. Downsize motors and correct problems at source. This is the most effective and positive long range approach however for expediency and avoiding immediate higher initial costs, adding capacitors can be of great value.

2. Utility company connects capacitors to high voltage lines before transformers and the meter outside the plant. If electrical load is not uniform day and night, switch gear is required to engage and disengage capacitors.

    On line installation runs about $15 per KVAR.

    Disadvantages are that this does nothing for in plant problem and doesn't register on meter.

3. Install capacitors on plant side of the transformer and the meter on the plant low voltage circuit.

    Correction is now read by meter but switching gear is still needed and this location does not correct problem at upstream sources.

    This one line installation runs about $16.50 per KVAR.

4. Best location for capacitors is at the source of the problem, the under loaded motors.

    In this case mount capacitor enclosure near starter and connect with 3 leads to starter terminals.

    A bonus feature of mounting the capacitors at the source in the correction of I^2R internal losses and improved voltage regulation on the feeders.

    The purchase price of the capacitors are from $10 to $12 per KVAR for 480 volts, 3 phase circuits in 10 to 100 KVAR sizes and with integral resistors and fuses various manufacturers are GE, Westinghouse, Federal Pacific, Sprague and others.

5. Do nothing and continue to pay additional power factor penalty charges.

Where:
KVAR = Kilovolts Amperes Reactive
(Measure of size of capacitors)

Figure 22-8
Wiring and Conduit
Per linear foot for 3 wires, furnished and installed 12 feet high

| Amps | Wire Stranded Thw, Copper Conduit Price | | Conduit Rigid Galv. | | Labor Hrs/Lf | Total Installed |
	Size	Cost 3 Wires	Dia.	Cost Lf	& 3 Wires	With 30% O&P
15	#14	$0.21	1/2 "	$1.02	0.070	$5.96
20	12	0.28	1/2	1.02	0.074	6.30
30	10	0.41	1/2	1.02	0.082	6.98
45	8	0.72	3/4	1.39	0.100	8.98
65	6	0.94	1	1.93	0.120	11.22
85	4	1.41	1	1.93	0.130	12.45
100	3	1.77	1-1/4	2.53	0.146	14.69
115	2	2.11	1-1/4	2.53	0.149	15.33
130	1	2.98	1-1/4	2.53	0.152	16.64
150	1/0	3.63	1-1/2	2.98	0.184	20.07
175	2/0	4.50	1-1/2	2 98	0.194	21.83
200	3/0	5.53	2	4.07	0.204	25.22

Labor at $48.00 per hour; Selling price includes 30% markup on material
1.   Thin wall conduit versus rigid installed price 88

Figure 22-9
Motor and Starter Engineering Data
3 phase, 1800 RPM, induction 230/460V

| Hp | Full Load Amps | | | Starter Size | | Percent Efficiency |
	115V	230V	460V	230V	460V	
1/2	4	2	1.0	00	00	70
3/4	5.6	2.8	1.4	00	00	72
1	7.2	3.6	1.8	0	00	79
1-1/2	10.4	5.2	2.6	0	0	80
2	13.6	6.8	3.4	0	0	80
3	19.2	9.6	4.8	0	0	81
5	30.4	15.2	7.6	0	0	83
7-1/2		22	11	1	0	85
10		28	14	1	1	85
15		42	21	2	1	86
20		54	27	2	1	87
25		68	34	2	2	88
30		80	40	3	2	89
40		104	52	3	2	89
50		130	65	4	3	89
60		154	77	5	4	89
75		192	96	5	4	90
100		248	124	5	4	90

Rule of thumb of amps
1. Lower voltages draw higher amps. A 115V motor draws half the amps of a 230V.
2. Higher voltages draw lower amps. A 230V motor draws double a 460V.
3. Single phase motors draw double the amps of 3 phase motors.
4. Rough calculation of amps for higher horse powers.
   Amps = 2.64 × horse power for 230 volts
   Amps = 1.32 × horse power for 460 volts

LIGHTING & ELECTRICAL
Check Off List

Reduce Lighting Levels
— Reduce Watts Per Square Foot
   Commensurate With Actual Needs
— Remove Florescent Bulbs
— Disconnect Ballasts Where Tubes
   Have Been Removed
— Completely Disconnect Certain
   Fixtures

Starters
— Change Thermal Overloads
— Change Starter Size
— Change Wiring Size
— Disconnect Switches
— Change Fuse Sizes

Replacement & Maintenance
— Replace Incandescent bulbs with Florescent
— Install High Efficiency Lamps and Ballasts

Operation & Maintenance
— Turn Lights Off When Spaces are Unoccupied
— Clean Bulbs, Reflectors and Lenses Yearly

Efficient Lighting System Design
— Use Specific Task Lighting
— When Natural Light is Available in perimeter Areas,
   Turn Off Lights
— In High Ceiling Areas, Lower Light Fixtures
— Split Up Switching So One Control Does Not Operate
   All Lights in Larger Rooms

Outside Lighting
— Reduce Outside Lighting Commensurate With Safety
— Use photo-Cell Control On Outdoor Lights

Motors
— Smaller Motor
— Energy Efficient Motor
— Variable Speed Motor
— Change Size Slide Rail
— Change pulley
— Change Belt Size

This chapter covers the following:

METHODS OF RECOVERING WASTE HEAT

How to Use Recaptured Heat

1. Heat up makeup air or temper outside air for HVAC systems.

2. Preheat combustion air.

3. Preheat hot water for central heating coils or terminal reheat coils in HVAC units.

4. Preheat hot water for baseboard heating.

5. Use heat recovery air directly for space heating or mix with regular supply air.

6. Preheat domestic hot water.

Stack Heat Recovery

- Install heat recovery unit in boiler stacks.
- Install heat recovery unit in incinerator stacks.

Engine Exhausts Heat Recovery
• Capture heat from engine exhaust.

Hot Exhaust Air Heat Recovery
• Install heat recovery units in kitchen exhaust systems, over baking ovens.

• Capture heat from painting processes, paint spray booths, parts washers, parts dryers and paint baking ovens.

• Install heat recovery units in printing plants.

• Install heat recovery units in industrial chemical exhaust systems.

• Recover heat from industrial drying exhausts of grain, milk, coffee, etc.

• Recapture heat from ovens used for baking, drying, etc., from kilns.

• Install heat recovery units in heat treating plants and annealing furnaces.

• Capture heat from clothes dryers in laundries.

• Install heat recovery equipment in foundries.

Water Cooled Condenser Heat Recovery
1. Reclaim heat from chiller with double bundle condenser which would otherwise be dumped to the atmosphere via the cooling tower.

2. Recapture heat with twin tower enthalpy recovery system.

Air Cooled Condenser Heat Recovery
• Reclaim heat from hot condenser air from air cooled DX condenser.

• Reclaim heat from hot refrigerant gas with gas to air or gas to glycol water transfer in ice skating rinks, supermarkets, computer rooms etc. Use for hot water heating or space heating.

Recirculate Warm Air from Internal Heat Gains
• Recirculate heat from equipment in computer rooms.

- Recirculate heat from interior lighting fixtures to heat perimeter areas.

Steam
- Recover heat from steam condensate.

Domestic Hot Water Recovery
- Reclaim heat from domestic hot water.

Machinery Heat Recovery
- Reclaim heat from water used to cool machinery.

HEAT RECOVERY RULE OF THUMB SAVINGS

Boiler Stack
1. Savings per year for 1 million Btuh input to boiler.

   Amount of Btu's of heat reclaim per million Btuh input @ 75% combustion efficiency and at a reclaim efficiency of 50%.

   $$\text{Btu/yr Reduction} = \frac{250{,}000 \, \text{Btuh}}{2} \times 2000 \, \text{hr/yr}$$

   $$= 250 \, \text{mill Btu/hr}$$

   Energy Costs Savings Per Mill. Btuh Input
   $1250/yr svgs; fuel @ $5/mill Btu
   $1750/yr svgs; fuel @ $7/mill Btu

2. Heat Recovery and Temperature Drop for 5 mill Btuh input to boiler at 75 percent efficiency of combustion.

   $$DT = \frac{\text{Btu Heat Reclaim}}{1.08 \times 5400}$$

   $$= \frac{625{,}000}{1.08 \times 5400}$$

   $$DT = 116°F$$

3. Combustion Air Required for gas combustion

    a) 11 cu ft of air per cu ft of gas or per 1000 Btuh of gas
    b) 183 Cfm combustion air required for 1 mill Btuh input
    c) 915 Cfm combustion air required for 5 mill Btuh input

4. Water Required for Heat Recovery Unit 1 million Btu input to boiler, 75% efficiency, 50% heat reclaim

    $$Btuh = 500 \times Dt \times Gpm$$

    $$Gpm = \frac{Btuh}{500 \times DT}$$

    $$= \frac{625,000}{10,000}$$

    Gpm = 62.5 Gpm for 50% reclaim

5. Costs per year for 5 million Btuh input boiler
    $6250/yr; fuel @ $5/mill Btu
    $8750/yr; fuel @ $7/mill Btu

Figure 23-1
Ways to Recover Heat

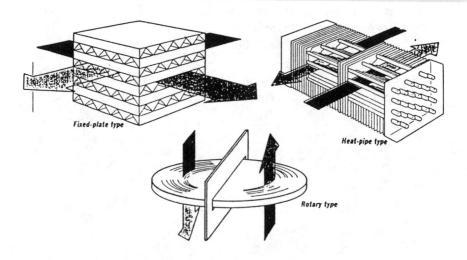

Fixed-plate type

Heat-pipe type

Rotary type

Figure 23-2
Heat Recovery Loop

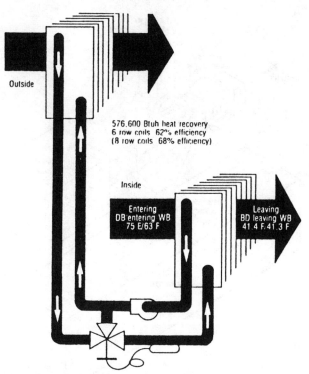

Outside

576,600 Btuh heat recovery
6 row coils 62% efficiency
(8 row coils 68% efficiency)

Inside

Entering
DB entering WB
75 E/63 F

Leaving
BD leaving WB
41.4 F.41.3 F

1 Heat recovery loops that preheat makeup air with exhaust air can cut costs in buildings
using large amounts of outside air.

Figure 23-3
Foundry Application

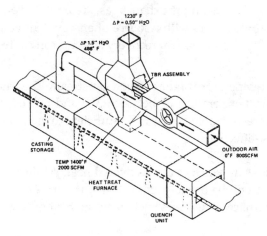

1230° F
ΔP = 0.50" H2O

ΔP 1.5" H2O
488° F

TBR ASSEMBLY

CASTING
STORAGE

TEMP 1400° F
2000 SCFM

OUTDOOR AIR
0° F 800SCFM

HEAT TREAT
FURNACE

QUENCH
UNIT

Figure 23-4
Incinerator Application

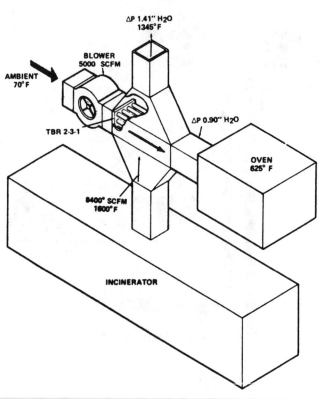

EVALUATING WASTE HEAT RECOVERY

Recovering heat from the exhaust gas stream of boilers and process furnaces can provide users with a source of energy that would otherwise be lost. In most applications, heat recovery systems can cut fuel costs by 10 to 20 percent.

1. Check the temperature of the exhaust gas stream. The gas should have a minimum temperature of between 350 and 500 degrees F, if the quantity of recoverable heat is to be sufficient to make a heat recovery project cost-effective.

2. Identify areas in which to use recovered heat. While most heat recovery systems will be use to preheat feedwater or combustion air for a boiler or furnace, other uses such as waste heat recovery boilers and

steam turbines for electric power generation can also be considered.

3. Be sure that the heat recovery system will be located close to the process where the captured heat will be used.

4. While no absolute rule is possible, boilers capable off generating 25,000 pounds per hour of steam and above are generally the best candidates for heat recovery. Boilers as low as 10,000 pph capacity should not, however, be ruled out.

5. Determine the number of hours that the boiler or furnace is used daily. Boilers and furnaces should be in operation a minimum of eight to 12 hours a day.

6. Make sure all metering and instrumentation on boilers or furnaces are well maintained and operating correctly.

7. Evaluate the system's operating efficiency. Any burner or boiler chosen for heat recovery should be well maintained and have a minimum efficiency of 60 to 70 percent.

8. Estimate a return on investment. In order to make a heat recovery project attractive, the return should be at least 10 to 15 percent higher than your cost of capital.

Table 23-1
Air-to-Air Plate Heat Exchangers

Cfm	Direct Material Cost		Labor	Total Material & Labor	
	Each	Per Cfm	Man Hours	Direct Costs	With 30% O&P
5,000	$7,590	$1.52	12	$8,034	$10,444
10,000	13,289	1.33	15	13,844	17,998
15,000	16,974	1.13	20	17,714	23,028
20,000	18,216	0.91	24	19,104	24,835
25,000	20,700	0.83	25	21,625	28,113

Accessories
1. Face and by-pass section with controls
    5,000 Cfm $1700
    10,000 Cfm $2000

Table 23-2
Heat Exchanger Wheels

Cfm	Direct Material Cost		Labor	Total Material & Labor	
	Each	Per Cfm	Man Hours	Direct Costs	With 30% O&P
Package Unit					
7,000	$48,300	$6.90	18	$48,966	$63,656
10,000	60,030	6.00	26	60,992	79,290
15,000	70,380	4.69	32	71,564	93,033
20,000	75,900	3.80	40	77,380	100,594
25,000	77,625	3.11	41	79,142	102,885
30,000	86,940	2.90	42	88,494	115,042
35,000	99,015	2.83	51	100,902	131,173
40,000	110,400	2.76	55	112,435	146,166
50,000	138,000	2.76	60	140,220	182,286
Wheel Only					
7,000	$11,454	$1.64	12	$11,898	$15,467
10,000	15,180	1.52	15	15,735	20,456
15,000	20,700	1.38	20	21,440	27,872
20,000	22,356	1.12	24	23,244	30,217
25,000	22,770	0.91	25	23,695	30,804
30,000	24,012	0.80	26	24,974	32,466
35,000	25,116	0.72	31	26,263	34,142
40,000	27,048	0.68	33	28,269	36,750
50,000	33,120	0.66	36	34,452	44,788

Package unit includes 2 fans, AC inverter on motor for wheel, all controls. Direct labor costs are $37.00 per hour.

Table 23-3
Exhaust Gas Heat Recovery Exchanger Shells
Carbon Steel

Dia. Inches	Length Inches	Direct Costs Each	Labor Man Hrs	Total Matl & Labor Direct Costs	With 30% O&P
4	26	2000.00	3.60	2097.20	2726.36
6	34	3000.00	7.50	3202.50	4163.25
8	42	4000.00	9.00	4243.00	5515.90
12	58	7000.00	11.40	7307.80	9500.14
16	74	11000.00	17.50	11472.50	14914.25
20	90	17000.00	21.00	17567.00	22837.10
24	106	25000.00	25.50	25688.50	33395.05
28	127	32000.00	32.00	32864.00	42723.20

Correction Factors
1. Stainless steel material, plus 50%
Heat taken from engine exhausts, etc.

Accessories
1. Manual Diverter $1,500 to $3,500
2. Automatic Diverter $1,865 to $4,000
3. Removable Manifold $400 to $1,000
Paybacks, 1 to 2 years

Uses of Recovered Heat
1. Heating air for HVAC systems.
2. Use in evaporation cycle of AC system.
3. Preheating feedwater to boilers.
Direct labor costs are $37.00 per hour.

Figure 23-5
Stack Heat Recovery

ECONOMIC EVALUATION OF AIR-TO-AIR ENERGY RECOVERY

In analyzing energy-recovery application, consideration should be given to first costs and operating costs.

First Costs, Expenditures
1. Costs of energy-recovery equipment
2. Installation labor of energy-recovery equipment
3. Additional ductwork required.
4. Larger fans or motors to overcome static pressure loss of energy-recovery equipment.
5. Additional air filtration required, if any.
6. Additional controls required.

First Costs Savings on Equipment Due to Reduced Loads
1. Boilers or other-heating equipment.
2. Heating coils and associated piping.
3. Cooling coils and associated piping.
4. Pumps.
5. Electrical equipment.

Extra Operating Costs
1. Maintaining the energy-recovery equipment.
2. Operating fans to overcome additional static pressure.
3. Maintaining additional filtration, if any.
4. Operating energy-recovery drive, pumps, controls, and defrost heaters.

Energy Savings
1. Savings of annual heating energy costs.
2. Savings of annual cooling energy costs.

RECOVER WASTE HEAT
Check Off List

Flue Gases; Stack Heat Recovery
— Boilers
— Engine Exhausts
— Incinerators

Hot Exhaust Air Recovery
— Kitchen Exhaust
— Baking Oven
— Drying Oven
— Laundry Dryers
— Drying Processes
— Kilns
— Printing Processes
— Swimming Pools
— Spray Painting Booths
— Heat Treating
— Foundries

Refrigerant Recovery
— Ice Skating Rinks
— Supermarkets
— Use Hot Refrigerant Gas
— Use Hot Condenser Air
— Computer Rooms
— Use For Hot Water Heating
— Use For Space Heating

Condenser Water
— Double Bundle Condenser

Steam
— Condensate Heat Recovery

Hot Water Recovery
— Domestic Waste Hot Water
— Hot Water From Machinery Water Cooling

Air-to-Air Exchangers
— Cross Flow Plate
— Fixed Plate Counter
— Flow
— High Temperature Plate
  Exchanger
— High Temperature Tube
  Exchanger
— Rotary Wheel
— Multiple Tower Exchanger

Air to Liquid Air
— Run Around Water Coils
— Thermosiphon

Liquid to Liquid
— Chiller Warm Condenser
  Water to Water

Hot Gas to Liquid
— DX to Water
— Stack Gas to Water

Dryer Cooler Exchange

Chapter 24

Testing, Balancing, Miscellaneous Costs

ESTIMATING AIR TESTING AND BALANCING WORK

	New System		Rebalance	
	Hr Range	Avg Cost w/ O & P	Hr Range	Avg Cost w/ O & P
Check Out and Test Equipment				
Built up Supply Unit	4.0 - 6.0	$200	3.0 - 5.0	$160
Air Handling Unit	3.0 - 5.0	160	2.0 - 4.0	120
Roof Top Unit, SZ	2.0 - 3.0	100	1.5 - 2.5	80
Roof Top Unit, MZ	2.0 - 4.0	120	1.5 - 3.0	100
plus per zone	.5	20	.4	16
Fan-Coil units under windows	0.5 - 1.0	30	0.5 - 1.0	30
Centrifugal Fans	2.0 - 3.0	100	1.0 - 2.0	60
Roof Exhaust Fans	1.0 - 2.0	60	0.5 - 1.0	30
Dust Collectors	2.0 - 4.0	120	1.0 - 3.0	80
Balance Outlets (No equipment time included)				
Supply Diffusers: Flow Hood	.25 -.35	12	.20 -.25	10
Other	.35 -.50	17	.25 -.35	12
Linear Diff.,Per 4ft: Flow Hood	.25 -.35	12	—	10
Air/Light Troffers: Flow Hood	.25 -.35	12	—	10
Small Registers: Flow Hood	.20 -.30	10	—	8
Other	.30 -.40	14	—	12
Large Registers: Flow Hood	.25 -.35	12	—	12
Other	.35 -.50	17	—	17

(Continued)

New System	Rebalance			
	Hr Range	Avg Cost w/ O & P	Hr Range	Avg Cost w/ O & P
Small Exhaust Hoods:	0.4 - 0.6	20	—	—
Large Exhaust Hoods:	0.6 - 0.8	28	—	—

Air Terminal Units, CAV and VAV

Above Ceiling:	0.4 - 0.6	20	—	—
(Check every 3rd unit)				
Balance Window Induction Units:	0.5 - 0.724	—	—	

Labor Correction Factors (Multipliers)
1.   Small, simple systems, open areas 0.75
2.   Large, complicated systems, many separate areas 1.25
NOTES
1.   Labor costs with overhead and profit, $40.00 per hour.
2.   Rebalance time, 60 to 80 percent of new balance time.
3.   Abbreviations; SZ: Single Zone MZ: Multi-Zone Other: Other type instrument, anomometer, hot wire, etc.

Figure 24-2. Estimating Ductwork Leak Testing
Medium Or High Pressure Duct Runs

LABOR FOR TYPICAL DUCT RUN

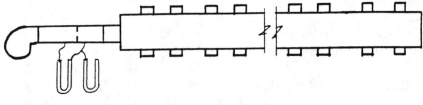

1.   Set up leak testing rig   1.0 hr
2.   Cap and seal ends of duct run, 1/2 hr each   1.0 hr
3.   Cap and seal branch collars, 16 @.2 hr each   3.2 hr
4.   Check leakage with fan, walk run, seal   2.0 hr
5.   Retest   1.0 hr
                    Total time for segment   8.2 hrs.

Figure 24-3
Estimating Hydronic Testing and Balancing

Check Out Equipment

                                                                    Hours

Pumps:	Check out pump itself, motor, starter, adjacent valves; read pressures, gpm, amps, volts; adjust.	1.5 to 2.5
	Chillers, Absorption Units	2.0 to 3.0
	Cooling Tower	5.0 to 10.0
	Central Cooling and Heating Coils	2.0 to 4.0

Balancing Terminals
Reheat Coils, Radiation Units	.75 to 1.25
Induction Units, Fan Coil Units	.75 to 1.25

CRANES AND LIFTING EQUIPMENT RENTAL

Telescoping Hydro Cranes
Range in capabilities between 4 and 30 tons.
15 ton unit lifts 3000 lbs 70 foot high at 30°, 100 high with a job.

15 ton crane most commonly used ....1/2 day $480
full day $672

Crawler Cranes, Lattice Type
Have to be assembled on the job site. Takes 3 men to assemble and 3 to disassemble.

Total costs runs the price of the labor plus $740 to $903/day rental.

Linden Tower Cranes
Counter balanced, rotating crane raised up with the building as it is erected. It sits in the middle and swings around as needed.

Common rental fee from general contractor $162 to $192/hr

Tower Hoist Rental - $80 per hour

Helicopters
Helicopter that will lift about 5000 pounds; may run around $1,110 base fee plus $112 per lift. Takes about half an hour to sling and hoist each piece as an average.

Scissors Hoists
Manually propelled scissors hoist, 18 foot platform height.
2-1/2 × 6-1/2 foot platform.

Rent ....$368 week $1,968/mo.

Self propelled, gas powered scissors hoist, 2000 lb capacity, 5'8" × 13'5" platform.

Platform height 25 feet   Rent ....$687 week

.................................$2,061/mo.

Forklift Trucks
Truck with 15 foot lift rents for $74/day

GENERAL CONSTRUCTION EQUIPMENT RENTAL

Abrasive Hole Cutting Saws
Rents for $46/day or $116/week

Backhoes
Backhoes for excavating rent for $236/day or $711/week

Structural Support on Roof for HVAC Equipment
For a 15 foot span between 2 joists or beams, an 8 inch "I" beam, running 15.3 lbs per linear foot will safely support a small or moderate size air conditioner, fan or condenser.

Concrete Pads
Small Pads, 4 × 4 feet, 6" high, 1/2 cubic yard

Singles	$430 minimum
Multiples	$364/cu yd

Large Pads, 20 × 10 feet, 6" high, 3 cubic yards
    Singles $430 minimum
    Multiples $310/cu yd

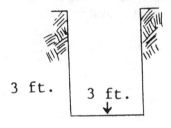

Excavating and Backfilling

3 ft × 4 ft deep trench, machine excavated backfilled and tamped, runs around $24.84/cu yd for regular soil.
Sandy soil is 25% less, clay soil 50% more.

3 ft.   3 ft.

Removal Work

Ductwork removal is about 50%, in existing buildings, of what the new installation rate is.
Equipment removal is 33% of new installation rate.

EXISTING BUILDING CONSIDERATIONS

Working on HVAC systems in existing buildings takes more time than it does in new buildings. It involves more complications and requires that many additional factors be considered when estimating or scheduling than as with new building work.

Openings to Cut	Do openings need to be cut?
	Do they have to be patched?
	What's the material and thickness
	of walls, roofs or floors to be cut?
	Quantity and sizes of openings?
Ceilings to Remove	Do ceilings have to be removed?
	Who replaces ceilings?

	Entire ceilings being removed or just cut where needed? What type of ceilings are involved and how high are they?
Removal of Existing HVAC Equipment, Etc.	Fans, pumps, HVAC units, ductwork, piping, etc?
Protection	Does owners furniture or equipment have to be protected?
Ceiling Space Available	What ceiling space is available? Where are beams located?
Equipment Room Space Available	Is there space for the new HVAC equipment, ductwork, piping, etc?
Obstructions To Avoid	Piping and ducts Lights and conduit Walls and partitions Beams, joists and columns
Occupied Areas	Will areas be occupied during installation?
Shutting Down of Systems	Must systems be in operation or can they be shut down during installation?
Sequence of Work	In what order must work be done?
Working Times	Week days, evenings, weekends. Over time
Material Handing And Hoisting	Elevators available Dock available Door, window, wall openings available and large enough for moving equipment through
Clean up and Scraping	Material handling labor, dumpster rentals, scavenger services.

DRAFTING LABOR

Fully detailed, 1/4 inch scale, office prepared shop drawings with locations, elevations, outlet locations. fitting details, pipe and fitting lengths, walls, partitions, and reflected lights and beams all shown.

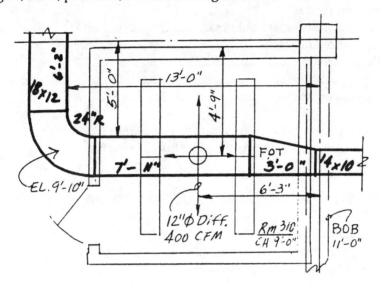

Labor

Labor includes the preparation of the shop drawings, revisions, field checks, taking off shop fabrication tickets, and listing blankouts.

It is a function of the quantity of pieces of ductwork in an area rather than the weight, fittings taking twice the total amount of time,from beginning to end,than straight pipe sections. It is also dependent on conjection in the ceiling space of all the mechanical, electrical work, on the complexity of the duct runs, and the extent of the architectural, structural involvement in the particular area.

Straight pipe ....1 HR/PC
Fittings ....2 HR/PC
Typical 50/50 mix, P & F by pieces, ...15 HR/PC

On a pounds per hour basis for an average mix of gauges:

10-20% fittings by wt	235 lbs/hr	Pipe only 350 lbs/hr
20-30% fittings by wt	200 lbs/hr	Fitt. only 100 lbs/hr
30-40% fittings by wt	185 lbs/hr	
40-50% fittings by wt	175 lbs/hr	

Correction Factors

Clear, open areas or straight runs...7
Congested ceiling spaces 1.2
Equipment room 1.2
Heavy duplications...8
Complete duplication of area 6

| **Budget Figures** | Direct costs: | 14¢/lb or 16¢/sq ft |
| | Sell Price: | 20¢/lb or 23¢/sq ft |

ESTIMATING FIELD MEASURING AND SKETCHING

Final Duct Connections
to units, fans, louvers, flexible connections,   120 lbs/hr   .5 hr/pc
average 22 gauge, 48×24, 60 lbs per pc

Complete Duct Runs
Measure area, obstructions, sketch run, figure lengths, elevations, draw
fittings - 30% fittings by weight, avg 4 ft long, 30 lb pcs for 24 ga

0 - 800 lbs	24 gauge avg	120 lbs/hr	.25 hr/pc
	22 gauge avg	200 lbs/hr	.25 hr/pc.
800 lbs up	24 gauge avg	150 lbs/hr	.2 hr/pc
	22 gauge avg	300 lbs/hr	.2 hr/pc

Budget Figures for New Project
Use 5% of total metal weight for finals and for measuring sheet metal
specialities

eg: 50,000 lb job × 5% = 2500 lbs divided by 120 lbs/hr = 24 hrs total

Hence field measuring averages out to about 2000 lbs/hr or 2.5¢/lb based
on total weight of job.

SERVICE

Two methods of calculating service and punch list work:

A. 2 hours per piece of equipment
B. or 5% of the cost of heating, refrigeration, air handling equipment

END OF BID FACTORS

1. Sales Tax
    Required on material and equipment on private work which you are selling directly to a user.
    Not required on government projects, non profit organizations, jobs for resale.

2. Performance Bonds
    Current rate about 1% of bid price.

3. Permits
    Check local codes on heating, refrigeration, ventilation, air pollution.

4. Wage Increases
    Determine schedule, calculate span of your work, add for wage increases.

5. Material Price Increases
    Determine when contracts will be awarded, when equipment purchases will be made, what price trend is on raw materials, galvanized, fiberglass etc. and when you'll need them. Add percent increase if anticipated.

6. Travel Pay
    Include travel pay per mile for traveling to job site if outside boundaries of local union jurisdiction,as required by contract.

7. Room and Board
    If you have to send your people to the job to stay include required room and board costs.

8.   Clean up Charges by General Contractor
     Check specs on house cleaning charges, if pro rated, who,s re-
     sponsible. Determine if.G.C. is one who slaps subs with clean up
     charges.

9.   Alternates
     Make sure alternates are included, especially on government
     projects where your bid may be rejected due to bidding irregular-
     ity.

10. Unit Prices
     Base unit prices on small quantities, on no discount situations, on
     low productivity rates, on heavy overhead involvement.

11. Special Contingencies

TESTING, BALANCING, MISCELLANEOUS COSTS
Check Off List

Drawings, Cartage
— Cartage
— Shop Drawings
— Field Sketching

Testing, Balancing, Service
— Testing and Balancing
— Pressure Testing
— Monitoring
— Startup
— O and M Manuals
— Set Up and Clean Up
—

Rental Costs
— Crane Rental
— Scaffolding Rental

General Construction Work
— Sleeves, Chases
— Cut Openings
— Patch Openings

Removal Work
— HVAC Equipment
— Ductwork
— Piping
— Electrical
— Partitions
— Ceilings
— Protect Building Furnishings
   and Equipment
— Scrap Items
— Clean Up

— Remove and Replace Ceilings
— Plaster
— Paint
— Excavate
— Backfill

Subcontractors
— Sheet Metal
— Piping
— Temperature Control
— Electrical
— Insulation
— Painting

End of Bid Factors
— Sales Tax
— Permits
— Bonds
— Travel Pay
— Room and Board
— Wage Increases
— Material Price Increases

Section 15

Appendices
General Data

Appendix A

Charts and Formulas

ABBREVIATIONS AND SYMBOLS

A	Area, Square Feet
A	Amps
act	Actual
BHP	Break Horsepower
Btu	British Thermal Unit
Btuh	British Thermal Unit per Hour
Cfm	Cubic Feet per Minute
Cu	Cubic
Cu.ft.	Cubic Feet
CHW	Chilled Water
CHWS	Chilled Water Supply
CHWR	Chilled Water Return
D	Difference, Delta
DB	Dry Bulb Temperature
DIA, dia	Diameter
DP, ΔP	Difference in Pressure
Dt, Δ t	Difference in Temperature
E	Voltage
eff	Efficiency
F	Fahrenheit
fpm	Feet per Minute
ft	Feet
GPH	Gallons per Hour
GPM	Gallons per Minute
H	Head
Hf	Friction Head
HP	High Pressure System
Hs	Elevation Head
HW	Hot Water
HWS	Hot Water Supply

HWR	Hot Water Return
I	Amperes
KVA	Kilovolt-Amperes
kW	Kilowatts
LP	Low Pressure System
MP	Medium Pressure System
MZ	Multi-Zone
NPSH	Net Positive Suction Head
Ns	Specific Speed
OA	Outside Air
P	Pressure
Pa	Atmospheric or Absolute Pressure
PF	Power Factor
Pvp	Vapor Pressure
R	Rankine Temperature
R	Resistance, Ohms
RA	Return Air
RH	Relative Humidity
rpm	Revolutions per Minute
SP	Static Pressure, Inches Water Gauge
Sp gr	Specific Gravity
sq ft	Square Feet
SZ	Single Zone System
t	Temperature, degrees Fahrenheit or Celcius
T	Absolute Temperature, 460 + Degrees Fahrenheit
TP	Total Pressure, Inches Water
V	Velocity, Feet per Minute
V	Volts
VP	Velocity Pressure, Inches Water Gauge
WB	Wet Bulb Temperature
WG	Water Gauge

Appendix B

Association Abbreviations

ACCA	Air Condition Contractors Association of America
ADC	Air Diffusion Council
ADI	Air Distribution Institute
AMCA	Air Moving and Control Association, Inc.
ARI	Air Conditioning and Refrigeration Institute
ASHRAE	American Society of Heating, Refrigeration and Air Conditioning Engineers, Inc.
ASME	American Society of Mechanical Engineers
MCAA	Mechanical Contractors Association of America, Inc.
NAPHCC	National Association of Plumbing-Heating Cooling Contractors
NEMA	National Electrical Manufacturers Association
NSPE	National Society of Professional Engineers
SMACNA	Sheetmetal and Air Conditioning Contractors National Association, Inc.

Appendix C
Conversion Factors

MULTIPLY	BY	TO OBTAIN
Atmospheres (Std.)		
760 MM of Mercury at 32° F.		
Atmospheres	14.696	Lbs./sq. inch
Atmospheres	76.0	Cms. of mercury
Atmospheres	29.92	In. of mercury
Atmospheres	33.90	Feet of water
Atmospheres	1.0333	Kgs./sq.cm.
Atmospheres	14.70	Lbs./sq. inch
Atmospheres	1.058	Tons/sq. ft.
Brit. Therm. Units	0.2520	Kilogram-calories
Brit. Therm. Units	777.5	Foot-lbs.
Brit. Therm. Units	0.000393	Horse-power-hrs.
Brit. Therm. Units	0.293	Watt-hrs.
BTU/min.	12.96	Foot-lbs./sec.
BTU/min.	0.02356	Horse-power
BTU/min.	0.01757	Kilowatts
BTU/min.	17.57	Watts
Calorie	0.003968	BTU

MULTIPLY	BY	TO OBTAIN
Feet of water	0.02950	Atmospheres
Feet of water	0.8826	Inches of mercury
Feet of water	0.03048	Kgs./sq. cm.
Feet of water	62.43	Lbs./sq. ft.
Feet of water	0.4335	Lbs./sq. inch
Feet/min.	0.5080	Centimeters/sec.
Feet/min.	0.01667	Feet/sec.
Feet/min.	0.01829	Kilometers/hr.
Feet/min.	0.3048	Meters/min.
Feet/min.	0.01136	Miles/hr.
Foot-pounds	0.001286	BTU
Gallons	3785	Cu. centimeters
Gallons	0.1337	Cubic feet
Gallons	231	Cubic inches
Gallons	128	Fluid ounces
Gallons	3.785	Liters
Gallons water	8.35	Lbs. water @60° F.
Horse-power	42.44	BTU/min.
Horse-power	33,000	Foot-lbs./min.
Horse-power	550	Foot-lbs./sec.
Horse-power	0.7457	Kilowatts
Horse-power	745.7	Watts
Horse-power (boiler)	33,479	BTU/hr.

MULTIPLY	BY	TO OBTAIN
Liters	0.2642	Gallons
Liters	2.113	Pints (liq.)
Liters	1.057	Quarts (liq.)
Meters	100	Centimeters
Meters	3.281	Feet
Meters	39.37	Inches
Meters	1000	Millimeters
Meters	1.094	Yards
Ounces (fluid)	1.805	Cubic inches
Ounces (fluid)	0.02957	Liters
Ounces/sq. inch	0.0625	Lbs./sq. inch
Ounces/sq. inch	1.73	Inches of water
Pints	0.4732	Liter
Pounds (avoir.)	16	Ounces
Pounds of water	0.01602	Cubic feet
Pounds of water	27.68	Cubic inches
Pounds of water	0.1198	Gallons
Pounds/sq. foot	0.01602	Feet of water
Pounds/sq. foot	0.006945	Pounds/sq. inch
Pounds/sq. inch	0.06804	Atmospheres
Pounds/sq. inch	2.307	Feet of water
Pounds/sq. inch	2.036	In. of mercury

Unit	Factor	Equals
Centimeters	0.3937	Inches
Centimeters	0.03280	Feet
Centimeters	0.01	Meters
Centimeters	10	Millimeters
Centimtrs. of Merc.	0.01316	Atmospheres
Centimtrs. of merc.	0.4461	Feet of water
Centimtrs. of merc.	136.0	Kgs./sq. meter
Centimtrs. of merc.	27.85	Lbs./sq. ft.
Centimtrs. of merc.	0.1934	Lbs./sq. inch
Cubic feet	2.832×10^4	Cubic cms.
Cubic feet	1728	Cubic inches
Cubic feet	0.02832	Cubic meters
Cubic feet	0.03704	Cubic yards
Cubic feet	7.48052	Gallons U.S.
Cubic feet/minute	472.0	Cubic cms., sec.
Cubic feet/minute	0.1247	Gallons/sec.
Cubic foot water	62.4	Pounds @ 60°F.
Feet	30.48	Centimeters
Feet	12	Inches
Feet	0.3048	Meters
Feet	1/3	Yards
Horse-power (boiler)	9.803	Kilowatts
Horse-power-hours	2547	BTU
Horse-power-hours	0.7457	Kilowatt-hours
Inches	2,540	Centimeters
Inches	25.4	Millimeters
Inches	0.0254	Meters
Inches	0.0833	Foot
Inches of mercury	0.03342	Atmospheres
Inches of mercury	1.133	Feet of water
Inches of mercury	13.57	Inches of water
Inches of mercury	70.73	Lbs./sq. ft.
Inches of mercury	0.4912	Lbs./sq. inch
Inches of water	0.002458	Atmospheres
Inches of water	0.07355	In. of mercury
Inches of water	0.5781	Ounces/sq. inch
Inches of water	5.202	Lbs./sq. foot
Inches of water	0.03613	Lbs./sq. inch
Kilowatts	56.92	BTU/min.
Kilowatts	1.341	Horse-power
Kilowatts	1000	Watts
Kilowatt-hours	3415	BTU
Pounds/sq. inch	27.68	Inches of water
Temp.(°C.)+273	1	Abs. temp. (°C.)
Temp.(°C.)÷17.78	1.8	Temp. (°F.)
Temp.(F.)÷460	1	Abs. temp. (°F.)
Temp.(F.)−32	5/9	Temp.(C.)
Therm	100,000	BTU
Tons(long)	2240	Pounds
Ton, Refrigeration	12,000	BTU/hr.
Tons (short)	2000	Pounds
Watts	3.415	BTU
Watts	0.05692	BTU/min.
Watts	44.26	Foot-pounds/min.
Watts	0.7376	Foot-pounds/sec.
Watts	0.001341	Horse-power
Watts	0.001	Kilowatts
Watt-hours	3.415	BTU/hr.
Watt-hours	2655	Foot-pounds
Watt-hours	0.001341	Horse-power hrs.
Watt-hours	0.001	Kilowatt-hours

Appendix D

Fuel Heating Values

Fuel	Heating Value

Coal

anthracite	13,900 Btu/lb
bituminous	14,000 Btu/lb
sub-bituminous	12,600 Btu/lb
lignite	11,000 Btu/lb

Heavy Fuel Oils and Middle Distillates

kerosene	134,000 Btu/gallon
No. 2 burner fuel oil	140,000 Btu/gallon
No. 4 heavy fuel oil	144,000 Btu/gallon
No. 5 heavy fuel oil	150,000 Btu/gallon
No. 6 heavy fuel oil, 2.7 % sulfur	152,000 Btu/gallon
No. 6 heavy fuel oil, 0.3% sulfur	143,800 Btu/gallon

Gas

natural	1,000 Btu/cu ft
liquefied butane	103,300 Btu/gallon
liquefied propane	91,600 Btu/gallon

Source: *Brick and Clay Record*, October 1972; reprinted with permission of the Cahner's Publishing Co. Chicago, Ill.

Appendix E

Air Heat Transfer Formulas

- Sensible

Btuh = Cfm × temp change ×1.0

Rearranged

$$Cfm = \frac{Btuh\ (Sensible)}{1.08 \times temp\ change}$$

Rearranged

$$Temp\ Change = \frac{Btuh\ (Sensible)}{Cfm \times 1.08}$$

Where

- Latent
  Btu = 4840   Xcfm × PH

- Total Latent and Sensible
  Btuh = 4.5 × Cfm × DH

Btuh	= British thermal units per Btuh hour
T	= Temperature, F
Cfm	= Cubic feet per minute
EFF	= Efficiency
HD	= Difference in enthalpy
WD	= Difference in humidity ratio (lb water/lb dry air)

STEAM FORMULAS

- Btuh

$$LB/HR = \frac{Btuh}{1000}$$

- Equivalent Direct Radiation

LB/HR = EDR ×.24

• Steam Coil In Duct

$$LB/HR = \frac{Cfm \times 1.08 \times TDa}{1000}$$

• Steam to Water Converter

$$LB/HR = gpm \times TDw \times .49$$

• CV of Steam Valve

$$Cv = \frac{Q}{63.5} \frac{V}{h}$$

Where
V   =   Specific Volume of Steam in cu.ft./lb.
h   =   Pressure Difference Across Valve
TD   =   Air Temperature Difference
TD   =   Water Temperature Difference
Specific Heat of Steam = .489 Btu/(LB) (F)

Appendix F

Air Flow and
Air Pressure Formulas

- Air Flow Formula

  Used to find volume of air flowing through ductwork, outlets, inlets, hoods., etc.

  Basic Formula $Cfm = A \times V$

  Velocity unknown: $V = \dfrac{Cfm}{A}$

  Area unknown: $A = \dfrac{Cfm}{V}$

  Where
Cfm	=	cubic fpm
A	=	area in sq ft
V	=	velocity in fpm
A_k	=	factor used with outlets; actual unobstructed air flow area

- Total Pressure Formula

  Measure of total pressure energy in air at any particular point in an air distribution system.

  $TP = VP + SP$

  rearranged

  $VP = TP - SP$

  $SP = TP - VP$

  where
TP	=	total pressure, inches W.G.

VP   =   velocity pressure, inches W.G.
SP   =   static pressure, inches W.G.

- Converting Velocity Pressure Into FPM

  Standard Air, 075 lb/cu. ft.

  $$fpm = 4005 \times \sqrt{VP}$$
  rearranged:

  $$VP = \left(\frac{FPM}{4005}\right)^2$$

  Non Standard Air

  $$fpm = 1096 \times \sqrt{\frac{VP}{density}}$$

where;
  VP   =   velocity pressure,
                 inches W.G.
  density   =   lb/cu ft
  fpm   =   feet per minute

Appendix G

Changing Fan
Cfm's and Driver

- Fan Law No. 1

$$\text{RPM new} = \text{RPM old} \times \left(\frac{\text{Cfm new}}{\text{Cfm old}}\right)$$

$$\text{SP new} = \text{SP old} \times \left(\frac{\text{Cfm new}}{\text{Cfm old}}\right)^2$$

$$\text{BHP new} = \text{BHP old} \times \left(\frac{\text{Cfm new}}{\text{Cfm old}}\right)^3$$

where

RPM	=	revolutions per minute
Cfm	=	cubic feet per minute
SP	=	fan static pressure, inches W.G.
BHP	=	break horsepower

- BHP Formulas

$$\frac{\text{BHP actual}}{\text{(3 phase)}} = \frac{1.73 \times \text{amps} \times \text{volts} \times \text{eff.} \times \text{power factor}}{746}$$

$$\frac{\text{BHP actual}}{\text{(rule of thumb)}} = \frac{\text{(name plate)}}{\text{horse power}} \times \left(\frac{\text{amps act.}}{\text{amps rated}}\right) \times \left(\frac{\text{volts act.}}{\text{volts rated}}\right)$$

where
BHP	=	break horsepower
eff	=	efficiency

- Sheave/RPM Ratios & Belt Lengths

$$\frac{\text{RPM motor}}{\text{RPM fan}}\left(\frac{\text{Speed}}{\text{Ratio}}\right)=\frac{\text{DIA fan Sheave}}{\text{DIA motor sheave}}\left(\frac{\text{Diameter}}{\text{Ratio}}\right)$$

$$\text{DIA fan sheave}=\text{DIA motor sheave}\times\left(\frac{\text{RPM motor}}{\text{RPM fan}}\right)$$

$$\text{DIA motor sheave}=\text{DIA fan sheave}\times\left(\frac{\text{RPM fan}}{\text{RPM motor}}\right)$$

$$\text{Belt Length}=2c+\left[1.57\times(D+d)\right]+\frac{(D-d)^2}{4c}$$

where

C	= center to center distance of shaft
D	= large sheave diameter
d	= small sheave diameter

Appendix H

Hydronic Formulas

- Converting PSI To Feet of Head

    ft hd = 2.31 × psi   inches hd = 27.2 × psi

    psi = .433 × ft hd,   psi = .036 × inches hd

- Water Heat Transfer Formulas

$$Btuh = GPM \times \left(T_{in} - T_{out}\right) \times 500$$

$$GPM = \frac{Btuh}{\left(T_{in} - T_{out}\right) \times 500}$$

- Electrical Power Consumption of Water Pump

$$BHP = \frac{GPM \times ft\ head}{3960 \times eff.}$$

- Using System Component as Flow Measuring Device

$$GPM\ actual = GPM\ design \times \sqrt{\frac{\Delta P\ actual}{\Delta P\ design}}$$

$$\Delta P\ actual = \Delta P\ design \times \left(\frac{GPM\ actual}{GPM\ design}\right)^2$$

- Coil or Chiller GPM

$$GPM = \frac{Tons \times 24}{T_{in} - T_{out}}$$

- Condenser GPM

$$GPM = \frac{Tons \times (kW \times .284)}{T\,out - T\,in}$$

where

GPM	=	gallons per minute
ΔP	=	change in pressure across component
Btuh	=	British thermal units per hour
T	=	temperature °F
BHP	=	break horsepower
kW	=	Kilowatts

Appendix I

Electrical Formulas

- **Three Phase, Alternating Current Motors**

$$\text{kW Actual (Motor Input)} = \frac{1.73 \times I \times E \times PF}{1000}$$

$$\text{BHP (Motor Output)} = \frac{1.73 \times I \times E - effM \times PF}{746}$$

$$\text{kWh (Motor Input)} = \frac{BHP}{effM}$$

- **If AMPS are unknown**

$$\text{AMPS (HP Known)} = \frac{BHP \times 746}{1.73 \times E \times effM \times PF}$$

$$\text{AMPS (kWh Known} = \frac{kWh \times 1000}{1.73 \times E \times PF}$$

$$\text{AMPS (KVA Known)} = \frac{KVA \times 1000}{1.73 \times E}$$

- **Kilo Volt AMPS**

$$KVA = \frac{1.73 \times I \times E}{1000}$$

- **Motor Electrical Costs For Year**

$$\text{Paid kWh Input Per year} = \frac{1.73 \times I \times E \times HR}{1000}$$

$$\text{kW Cost Per Year} = \frac{1.73 \times I \times E \times HR \times \$kWh}{1000}$$

Where

I	=	Current in amps
kWh	=	Kilowatt hours
E	=	Voltage
BHP	=	Break horsepower
PF	=	Power factor
KVA	=	Kilovolt amps
effM	=	Efficiency of motor
HR	=	Hour per year

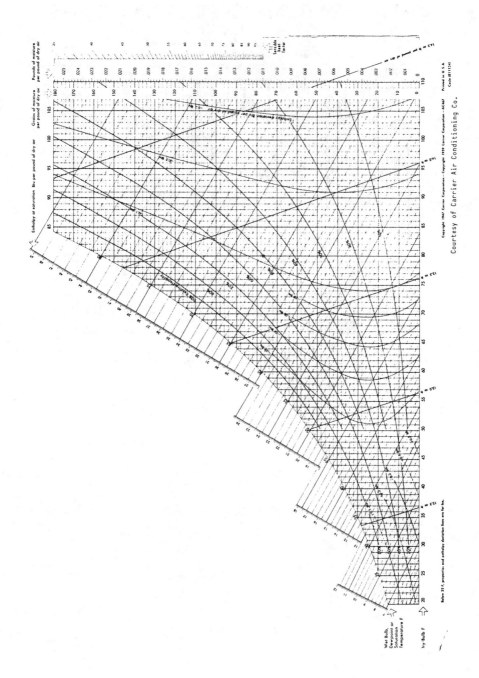

Courtesy of Carrier Air Conditioning Co.

REFRIGERATION AND AIR CONDITIONING

TEMPERATURE-PRESSURE CHART
Shaded Figures = Vacuum • Solid Figures = Pressure

°F	R-12	R-13	R-22	R-500	R-502	R-717 Ammonia	°F	R-12	R-13	R-22	R-500	R-502	R-717 Ammonia
-100	27.0	7.5	25.0	26.4	23.3	27.4	16	18.4	211.9	38.7	24.1	47.7	29.4
-95	26.4	10.9	24.1	25.7	22.1	26.8	18	19.7	218.8	40.9	25.7	50.1	31.4
-90	25.8	14.2	23.0	24.9	20.7	26.1	20	21.0	225.7	43.0	27.3	52.5	33.5
-85	25.0	18.2	21.7	24.0	19.0	25.3	22	22.4	233.0	45.3	28.9	54.9	35.7
-80	24.1	22.3	20.2	22.9	17.1	24.3	24	23.9	240.3	47.6	30.6	57.4	37.9
-75	23.0	27.1	18.5	21.7	15.0	23.2	26	25.4	247.8	49.9	32.4	60.0	40.2
-70	21.9	32.0	16.6	20.3	12.6	21.9	28	26.9	255.5	52.4	34.2	62.7	42.6
-65	20.5	37.7	14.4	18.8	10.0	20.4	30	28.5	263.2	54.9	36.0	65.4	45.0
-60	19.0	43.5	12.0	17.0	7.0	18.6	32	30.1	271.3	57.5	37.9	68.2	47.6
-55	17.3	50.0	9.2	15.0	3.6	16.6	34	31.7	279.5	60.1	39.9	71.1	50.2
-50	15.4	57.0	6.2	12.8	0.0	14.3	36	33.4	287.8	62.8	41.9	74.1	52.9
-45	13.3	64.6	2.7	10.4	2.1	11.7	38	35.2	296.3	65.6	43.9	77.1	55.7
-40	11.0	72.7	0.5	7.6	4.3	8.7	40	37.0	304.9	68.5	46.1	80.2	58.6
-35	8.4	81.5	2.6	4.6	6.7	5.4	45	41.7	327.5	76.0	51.8	88.3	66.3
-30	5.5	90.9	4.9	1.2	9.4	1.6	50	46.7	351.2	84.0	57.6	98.9	74.5
-28	4.3	94.9	5.9	0.1	10.5	0.0	55	52.0	376.1	92.6	63.9	108.0	83.4
-26	3.0	98.9	6.9	0.9	11.7	0.8	60	57.7	402.3	101.6	70.6	118.6	92.9
-24	1.6	103.0	7.9	1.8	13.0	1.7	65	63.8	429.6	111.2	77.8	125.8	103.1
-22	0.3	107.3	9.0	2.4	14.2	2.6	70	70.2	458.7	121.4	85.4	136.6	114.1
-20	0.6	111.7	10.2	3.2	15.5	3.6	75	77.0	489.0	132.2	93.5	148.0	125.8
-18	1.3	116.2	11.3	4.1	16.9	4.6	80	84.2	520.8	143.6	102.0	159.9	138.3
-16	2.1	120.8	12.5	5.0	18.3	5.6	85	91.8	–	155.7	111.0	172.5	151.7
-14	2.8	125.7	13.8	5.9	19.7	6.7	90	99.8	–	168.4	120.6	185.9	165.9
-12	3.7	130.8	15.1	6.8	21.2	7.9	95	108.3	–	181.8	130.6	199.7	181.1
-10	4.5	135.4	16.5	7.8	22.8	9.0	100	117.2	–	195.9	141.2	214.4	197.2
-8	5.4	140.5	17.9	8.8	24.4	10.3	105	126.6	–	210.8	152.4	229.7	214.2
-6	6.3	145.7	19.3	9.9	26.0	11.6	110	136.4	–	228.4	164.1	245.8	232.3
-4	7.2	151.1	20.8	11.0	27.7	12.9	115	146.8	–	242.7	176.5	262.6	251.5
-2	8.2	156.5	22.4	12.1	29.4	14.3	120	157.7	–	259.9	189.4	280.3	271.7
0	9.2	162.1	24.0	13.3	31.2	15.7	125	169.1	–	277.9	203.0	298.7	293.1
2	10.2	167.9	25.8	14.5	33.1	17.2	130	181.0	–	298.8	217.2	318.0	315.0
4	11.2	173.7	27.3	15.7	35.0	18.8	135	193.5	–	316.6	232.1	338.1	335.0
6	12.3	179.8	29.1	17.0	37.0	20.4	140	206.6	–	337.0	247.7	369.1	388.0
8	13.5	185.9	30.9	18.4	39.0	22.1	145	220.3	–	368.9	286.1	381.1	390.0
10	14.6	192.1	32.8	19.7	41.1	23.8	150	234.6	–	381.5	281.1	403.9	420.0
12	15.8	198.6	34.7	21.2	43.2	25.6	155	249.5	–	405.1	296.9	427.8	460.0
14	17.1	205.2	36.7	22.6	45.6	27.5	160	265.1	–	429.6	317.4	452.6	490.0

PROPERTIES OF SATURATED STEAM

Vacuum, Inches of Mercury	Boiling Point or Steam Temperature, Deg. F.	Specific Volume (V), cu. ft/lb	$\sqrt{v}$	Maximum Allowable Pressure Drop, psi. (For valve sizing)	Heat of the Liquid, Btu	Latent Heat of Evap., Btu	Total Heat of Steam, Btu
29	76.6	706.00	26.57	0.28	44.7	1048.6	1093.3
25	133.2	145.00	12.04	1.2	101.1	1017.0	1118.1
20	161.2	75.20	8.672	2.4	129.1	1001.0	1130.1
15	178.9	51.30	7.162	3.7	146.8	990.6	1137.4
14	181.8	48.30	6.950	3.9	149.7	988.8	1138.5
12	187.2	43.27	6.676	4.4	155.1	985.6	1140.7
10	192.2	39.16	6.257	4.9	160.1	982.6	1142.7
8	196.7	35.81	5.984	5.4	164.7	980.0	1144.7
6	201.0	32.99	5.744	5.9	168.9	977.2	1146.1
4	204.8	30.62	5.533	6.4	172.8	974.8	1147.6
2	208.5	28.58	5.345	6.9	176.5	972.5	1149.0
Gage Pressure, psig							
0	212.0	26.79	5.175	7.4	180.0	970.4	1150.4
1	215.3	25.20	5.020	7.8	183.8	968.2	1151.5
2	218.5	23.78	4.876	8.4	186.6	966.2	1152.8
3	221.5	22.57	4.750	8.8	189.6	964.8	1153.9
4	224.4	21.40	4.626	9.4	192.5	962.4	1154.9
5	227.1	20.41	4.518	9.8	195.8	960.6	1155.9
6	229.8	19.45	4.410	10.4	198.0	958.8	1156.8
7	232.3	18.64	4.317	10.8	200.5	957.2	1157.7
8	234.8	17.85	4.225	11.4	203.0	955.5	1158.5
9	237.1	17.16	4.142	11.8	205.4	954.0	1159.4
10	239.4	16.49	4.061	12.4	207.7	952.5	1160.2
11	241.6	15.90	3.987	12.8	209.9	951.1	1161.0
12	243.7	15.35	3.918	13.4	212.1	949.7	1161.8
15	249.8	13.87	3.724	14.8	214.2	948.3	1162.5
20	258.8	12.00	3.464	17.4	227.4	939.5	1166.9
25	266.8	10.57	3.250	19.8	235.6	934.0	1169.6

(Continued)

30	274.0	9.463	3.076	22.4	248.0	928.9	1171.9
40	286.7	7.826	2.797	27.4	255.9	919.9	1175.8
50	297.7	6.682	2.585	32.4	267.1	911.9	1179.0
60	307.8	5.886	2.415	37.4	277.1	904.7	1181.8
70	316.0	5.182	2.276	42.4	286.1	898.0	1184.1
80	323.9	4.662	2.159	47.4	294.8	891.9	1186.2
90	331.2	4.289	2.059	52.4	301.9	886.1	1188.0
100	337.9	3.888	1.972	57.4	308.9	880.7	1189.6
120	350.0	3.337	1.826	67.4	321.7	870.7	1192.4
140	360.9	2.928	1.710	77.4	333.1	861.5	1194.6
160	370.6	2.602	1.613	87.4	343.4	853.0	1196.4
180	379.6	2.846	1.531	97.4	353.0	845.0	1198.0
200	387.8	2.134	1.461	107.4	361.8	837.5	1199.8
250	406.0	1.742	1.320	132.4	381.5	820.2	1201.7
300	421.8	1.472	1.213	157.4	398.6	804.6	1203.2
350	435.6	1.272	1.128	182.4	414.1	790.1	1204.2
400	448.1	1.120	1.058	207.4	428.0	776.6	1204.6
450	459.5	0.998	0.999	232.4	440.9	763.7	1204.6
500	470.0	0.900	0.948	257.4	452.9	751.4	1204.8
550	479.7	0.818	0.904	282.4	464.0	739.7	1203.7
600	488.8	0.749	0.865	307.4	474.6	728.4	1203.0
650	497.3	0.690	0.831	332.4	484.7	717.8	1202.0
700	505.4	0.639	0.799	357.4	494.8	706.5	1200.8
800	520.8	0.554	0.746	407.4	512.8	685.9	1198.2
900	533.9	0.488	0.699	457.4	529.0	666.0	1195.0
1000	546.3	0.435	0.659	507.4	544.2	647.2	1191.4

Index